建筑与都市系列丛书 | 世界建筑
Architecture and Urbanism Series | World Architecture

文筑国际 编译
Edited by CA-GROUP

Latvia: Architecture Unfolding

拉脱维亚：建筑表达

中国建筑工业出版社

图书在版编目（CIP）数据

拉脱维亚：建筑表达 = Latvia Architecture Unfolding：汉英对照 / 文筑国际 CA-GROUP 编译 . -- 北京：中国建筑工业出版社，2021.1
（建筑与都市系列丛书 . 世界建筑）
ISBN 978-7-112-25723-2

Ⅰ . ①拉… Ⅱ . ①文… Ⅲ . ①建筑艺术－介绍－拉脱维亚 －汉、英 Ⅳ . ① TU-865.11.7

中国版本图书馆 CIP 数据核字 (2020) 第 247371 号

责任编辑：毕凤鸣 刘文昕
版式设计：李梦迪
责任校对：芦欣甜

建筑与都市系列丛书｜世界建筑
Architecture and Urbanism Series ｜ World Architecture
拉脱维亚：建筑表达
Latvia: Architecture Unfolding
文筑国际　编译
Edited by CA-GROUP
*
中国建筑工业出版社出版、发行（北京海淀三里河路 9 号）
各地新华书店、建筑书店经销
北京雅昌艺术印刷有限公司　制版、印刷
*
开本：787 毫米×1092 毫米　1/16　印张：17½　字数：550 千字
2024 年 7 月第一版　　2024 年 7 月第一次印刷
定价：148.00 元
ISBN 978-7-112-25723-2
（36509）

a+u

建筑与都市系列丛书
Architecture and Urbanism Series

总策划 Production
国际建筑联盟 IAM　文筑国际 CA-GROUP

出品人 Publisher
马卫东 MA Weidong

总策划人/总监制 Executive Producer
马卫东 MA Weidong

内容担当 Editor in Charge
吴瑞香 WU Ruixiang

助理 Assistants
王霄晗 WANG Xiaohan　顾 焕 GU Huan　郝雅楠 Hao Ya'nan

翻译 Translators
中译英 English Translation from Chinese:
朱亦婷 ZHU Yiting (p275)
英译中 Chinese Translation from English
盛 洋 SHENG Yang (pp.12-19, 30-37, 84-93, 184-191, 246-253)
郝雅楠 HAO Ya'nan (pp.38-83, 94-133)
日译中 Chinese Translation from Japanese:
吴瑞香 WU Ruixiang (p15) 卢亭羽 LU Tingyu (p277)

书籍设计 Book Designer
李梦迪 LI Mengdi

中日邦交正常化50周年纪念项目
The 50th Anniversary of the Normalization of China-Japan Diplomatic Relations

本系列丛书部分内容选自A+U第555号（2016年12月号）特辑
原版书名：
ラトヴィア——建築の表出
Latvia - Architecture Unfolding
著作权归属A+U Publishing Co., Ltd. 2016

A+U Publishing Co., Ltd.
发行人/主编：吉田信之
客座主编：伊尔泽·帕克罗内
编辑：服部真吏
海外协助：侯 蕾

Part of this series is selected from the original a+u No. 555 (16:12),
the original title is:
ラトヴィア——建築の表出
Latvia - Architecture Unfolding
The copyright of this part is owned by A+U Publishing Co., Ltd. 2016

A+U Publishing Co., Ltd.
Publisher / Chief Editor: Nobuyuki Yoshida
Guest Editor: Ilze Paklone
Editorial Staff: Mari Hattori
Oversea Assistant：HOU Lei

封面图：城墙外的瑞典门
封底图：装饰艺术楼梯间改造，建筑设计：苏德拉巴建筑师事务所

本书在编辑过程中，部分信息的校对得到了拉脱维亚驻华大使馆的协助，在此表示特别感谢。本书第270页至279页内容由安藤忠雄建筑研究所提供，在此表示特别感谢。

Front cover: View of the Swedish Gate outside the city walls. Photo by Shinkenchiku-sha. Back cover: Staircase Renovation Art Deco, designed by Sudraba Arhitektura

During the editing process of this book, the proofreading of some information was assisted by the Latvian Embassy in China. We would like to express our special thanks here. The contents from pages 270 to 279 of this book were provided by Tadao Ando Architect & Associates. We would like to express our special thanks here.

Preface:

Stick to Cultural Specialty and Architectural Essence of Latvian-ness

HAN Linfei

Latvia, located on the eastern shore of the Baltic Sea, is one of the "Three Baltic States", together with Estonia and Lithuania. Since the establishment of the feudal duchy in the 13th century, she has been fighting for independence, freedom and liberation. Until the Soviet Union collapsed, Latvia regained its independence. By the historical perspective on Latvia, the particularity in urban construction is reflected in the diversity of architectural styles: from traditional farm and unique wooden buildings to the luxurious palaces and manor, from the great church to impressive art nouveau style building, from the majestic Stalinist buildings to the modernization construction. Although the urban buildings of different ages have distinct marks of the times, Latvians adhere to the local architecture and culture.

Latvia is known as the country of Art Nouveau architecture, and Riga is known as the capital of Art Nouveau architecture. On the street corners of Latvian cities, the elegant lines of geometric patterns and decorative relievo of old buildings represent almost the highest achievements of art Nouveau architecture in the world at the time. Moreover, the vast majority of Art Nouveau architecture was designed and built by Latvian architects, who led the trend of Art Nouveau architecture design in the 20th century.

In the Soviet era, Latvia's capital Riga, small towns and even the remote areas, have the obvious characteristics of the times. Especially, in Latvia there are many Stalinist public and residential buildings. At that time, although Latvia was greatly influenced politically, she still adhered to her own cultural cognition. For example, the Building of Latvian Academy of Sciences in Riga, is marked with the national form of Soviet socialist architectural content. However, the architects still insist on the architectural embodiment of the local cultural heritage in the design of its interior space and external environment. All of these demonstrate the persistence in her own national culture by Latvians in that period of history.

Since Latvia regained its independence in the 1990s, it has been engaged in national power recovery, economic recovery and urban construction for many years. A new generation of Latvians inherits the forefathers' traditional protection and restoration of historic buildings. The ancient buildings with special historical significance and cultural connotation such as Blackheads House were rebuilt. Some of the historic buildings were restored and modernized. The innovative design attempts by combining modern technology, material and local culture were made. The new buildings and ancient buildings organically integrated in the cities improve the urban environment and enhance the quality of civic life together.

The history of Latvian architecture is closely related to the history of national development. It is a history of struggle for independence, freedom and liberation. In Latvia stands many monuments and buildings symbolizing the spiritual strength of the people and the ideals of freedom and independence, showing Latvia's profound experience, historical and cultural characteristics. Although the Latvian architecture experienced a bumpy course, the Latvians have been sticking to the independent national spirit and cultural specialty. They are making localized design innovations and in the meanwhile preserving the historic buildings, constantly pursuing the traditional culture and texture in the architecture, finally developing the architectural essence of Latvian-ness seeking independence and freedom.

序言：

文化特殊性坚守
与建筑本质追求中的拉脱维亚性

韩林飞

拉脱维亚，地处波罗的海东岸，与爱沙尼亚和立陶宛并称为"波罗的海三国"。她是一个命运多舛的国度，自13世纪建立封建公国，一直在争取独立、自由和解放。历史上，她几度被他国占领，直到1991年苏联解体，重新独立建国。以历史的视角看待拉脱维亚，这种特殊性在城市建筑上表现为异彩纷呈的建筑风格：从传统的农场和独特的木构建筑到豪华的宫殿和庄园，从宏伟的教堂到印象深刻的新艺术风格建筑，从气势磅礴的斯大林式建筑到宜古宜今的现代化建筑。虽然承载着不同时代文化的城市建筑带有鲜明的时代印记，但是不变的却是拉脱维亚人对"本土建筑和文化之心"的坚守与追求。

拉脱维亚被称为"新艺术"建筑风格的国家，其首都里加更是享有"新艺术建筑之都"的美誉。那些位于拉脱维亚的各个城市的街角，古老华丽建筑上精细雅致的几何图案线条、装饰性的浮雕石像，几乎代表了当时全世界新艺术建筑风格的最高成就。而且绝大多数新艺术建筑风格的建筑为拉脱维亚建筑师设计建造，他们自强不息、止于至善，用才华引领20世纪的新艺术风格建筑设计潮流趋势。

历史的车轮滚滚向前，无论是在拉脱维亚的首都里加，还是在小城镇、甚至是边远地区，都可看到苏联时代的印记。最具代表性的是，在拉脱维亚，与苏联其他加盟共和国一样，屹立着许多斯大林式的公共建筑和住宅建筑。苏联时代，拉脱维亚虽然在政治上受到很大影响，但仍在坚守着自我的文化认知，坚持着本土文化的追求。如里加的拉脱维亚科学院大厦，虽带有苏联建筑的社会主义内容民族形式的明显印记，但其室内空间及外部环境的设计中，仍坚持着本土文化传承的建筑体现，展示了那段历史中对自身民族文化性的坚守。

拉脱维亚自20世纪90年代恢复独立以来，多年来一直致力于国力恢复、经济复苏和城市建设。新一代拉脱维亚人传承了先人们对历史建筑的保护与修复传统，对黑头宫这种具有特殊历史意义和文化内涵的古建筑进行重建，对部分历史建筑进行修复和现代化改造，同时将现代技术、材料与本土文化相结合，进行大胆的创新设计尝试，宜古宜今的新建筑与古建筑在城市中相得益彰，共同改善城市环境，提升公民生活品质。

拉脱维亚的建筑史与其国家发展史密不可分，是一部争取独立、自由和解放的奋斗史。这里屹立着许多象征人民精神力量和自由与独立理想的纪念碑与建筑，展示着拉脱维亚宽广而深厚的阅历和历史人文特征。尽管其建筑发展之路充满了艰辛与坎坷，但是拉脱维亚人一直秉承坚毅独立的民族精神和特殊的文化内核，在保护历史建筑的同时致力于本土化的设计创新，不断追求建筑中蕴含的传统文化与质感，发扬建筑本质中追求独立自由的拉脱维亚性。

HAN Linfei
Ph.D. of Architecture, Ph.D. of Urban Economics, Postdoctoral in Urban Geography, Professor, Moscow Architectural Institute Professor, Beijing Jiaotong University

韩林飞
建筑学博士，城市经济学博士，城市地理学博士后，莫斯科建筑学院教授，北京交通大学教授

Latvia: Architecture Unfolding

拉脱维亚：建筑表达

Editor's Words

编者的话

This is a special edition devoted to Latvian architecture, with a focus on 25 contemporary works completed since 2000, divided into four areas. At the same time it introduces the historical architecture that provides the context for these contemporary works. This approach follows naturally from the fact that many of the works introduced here evidence the greatest respect for historical architecture and the historical city. Especially in the Old Town of Riga, which is registered as a UNESCO World Heritage site, many harmonious renovations and reconstructions have taken place in which the architects conduct astoundingly strict and precise restorations and install the functions needed for modern life.

The task of defining a context implies the work of identifying the whole and clarifying its foundations. In the diverse and multi-layered physical and cultural context of Latvia, over the more than 30 years since the country achieved its independence, architects have demonstrated their answers to the question "What is Latvian architecture?" through their own work. This book is an attempt to interpret Latvian architecture through some of the representative works of contemporary Latvian architecture.
(a+u)

Translated from Japanese by Thomas Donahue

这是一本介绍拉脱维亚建筑的专辑，我们分四个区域，重点介绍了25座建于2000年以后的拉脱维亚当代建筑。同时也介绍了为这些当代建筑提供创作语境的历史建筑。从本书收录的诸多作品中，我们能感受到建筑师对历史遗存、对一座历史城市最大的敬意和关怀。特别是被联合国教科文组织列为世界遗产的里加老城，曾进行过多次更新与重建，建筑师们在极尽缜密修缮的同时，也配置了现代生活所需的功能，使之达到了古今调和的状态。

所谓定义语境，就是明晰何为整体、何为根本的工作。因此，建筑师们在解读国家在物质、文化层面多元、多层次的文明语境时，也通过他们自己的作品，自发地释明了他们一直以来所追索的问题——拉脱维亚独立30年以来，"什么是拉脱维亚建筑"？本书即试图通过当代拉脱维亚建筑作品来解读拉脱维亚建筑这一命题。

(a+u)

Editorial Essay:

Capturing the essence of 'Latvian-ness'

Ilze Paklone, Guest Editor

编辑论文：

捕捉"拉脱维亚性"的本质

伊尔泽·帕克罗内，客座主编

Being a Latvian architect in Tokyo for four years, I have been asked many times about what is Latvian-ness and what is special about culture, design and essentially architecture in Latvia. It has made me think about attempting to explain it, if not precisely, then figuratively through keywords, selected projects and comments from other architects. One way of explaining the concept of what can be termed Latvian-ness is through the written word. The same attitude, same abstraction of thought comes down to architecture. Latvian poems in the traditional four-line form of verses called dainas are compact and ascetic forms of thinking, taking great care with a figurative, harmonious and calm description of the world view that, perhaps, well describes the ascetic and laconic nature, materials and colors of Northeast Europe. Smallness and compactness of nature and things created, yet very rich aesthetic and spatial experiences, is what eventually is the closest to explain Latvian-ness. The other way to explain the Latvian attitude towards architecture is through relationships among people. Latvians are rather reserved, but at the same time this distance and silence is perceived as precious, mutually understood, calming and respectful. In addition to rich experiences of smallness and compactness, the silences, the pauses, distances, wideness and modesty are also closest to explaining Latvian-ness.

For this book it has been a very natural desire to trace architecture in Riga as the city has been a vibrant urban center for almost nine centuries, currently being home for more than one third of the country's population. What are the constituents that shape refinement of attitudes towards the city of Riga, to both physical and intangible legacies? In its essence, Riga has always captured the exact moment of transition from one cultural paradigm to the next one. One way to read it is to name this process as continuous unfolding and layering of cultural ideas, being sensible, juvenile and eagerly open to the new streams, at the same time always being respectful to the existent layers of history. Surprisingly, despite the small scale of the city, a surplus of the continuous foldings of culture in Riga is condensed in the metropolitan flavor of the city, taking part in the world's events. How then do today's architects see this condition of continuous unfolding in their own practices? The selected works in Riga represent this condition through multiple attitudes towards existent context, showcasing how gracefully multiple architectural styles, ideas and fragments blend and transit into each other. Another aspect that also shapes an important part of what could potentially be termed as contemporary Latvian-ness is a lifestyle that always fluctuates between urban and non-urban. The book also features the projects that serve as

作为一名在东京生活了四年的拉脱维亚建筑师，我曾多次被问及何为“拉脱维亚性”，拉脱维亚文化、设计以及根本上关于拉脱维亚建筑的特殊之处。这让我开始思考如何予以解释。要精确描述或许很难，但通过某些关键词、其他建筑师精选的一些项目和观点，我想或许我可以来做一种例证性的说明。书面语是解释“拉脱维亚性”的方式之一，文字传达的态度、思想，同样也会落在建筑上。拉脱维亚引以为珍的民谣“达伊纳斯”（拉脱维亚语为Dainas），采用四行诗的形式，短小精悍、禁欲地表达所思所想。“达伊纳斯”中往往对世界有着充满隐喻、祥和平静的描述，而这些似乎也在很好地表达东北欧克制凝练的风土自然、材料和色彩。无论天然还是人造，特有的“小”和“紧凑”都反而带来了丰富的审美体验和空间体验，最为接近“拉脱维亚性”的内涵。另一种解释拉脱维亚建筑观的方式是通过人际关系。拉脱维亚人其实相当保守，但与此同时，他们尤为珍视这份距离和沉默，因其意味着彼此间的理解，冷静和尊重。除了“小”和“紧凑”带来的丰富体验，寂静、沉默、距离感、广阔、谦逊亦是对“拉脱维亚性”最为贴切的解释。

里加在将近九个世纪里一直是一座活跃的中心城市，整个国家三分之一以上的人口都居住于此，追溯里加的建筑轨迹自然也成了这本书的目的之一。那么，如何理性地认识里加这座城市以及当地的物质或非物质遗产？本质上，里加总能捕捉到文化模式的次第变迁。我们可以将这种变迁解读为文化观念的展开和分层，换言之，既表现出对新事物理智、单纯和热切地接纳，同时保持对历史的尊重。你会惊讶地发现，尽管城市规模很小，里加层出不穷的文化里，凝聚着毫不逊色于世界各地的国际化都市气息。面对这种状态，今天的建筑师持有怎样的看法？针对这一问题，本书筛选了里加一系列有代表性的建筑项目。设计者对文脉的不同态度显示出，多种多样的建筑风格、理念和残垣断壁能优雅地融合或交叠。另一方面，介于城市和非城市之间的生活方式，也在很大程度上塑造了当代的“拉脱维亚性”，本书同样也关注此类“大隐于市”的项目。

《拉脱维亚：建筑表达》主要追溯了拉脱维亚自独立以来里加以及拉脱维亚的建筑图景。这段时间里，碎片化的理念在凝结，文化的进程和实践在加速，人们除了狂喜于1990年拉脱维亚重获独立，还迫切希望了解培养国家自主性的意义。本质而言就是这样的问题：什么让拉脱维亚超乎寻常？在更广阔的欧洲语境下，“拉脱维亚性”指代什么？拉脱维亚与未断代发展的国家，或与其他新生组织形态有何共通之处？过去的25年间，从被民间简化的碎片化的历史，到振奋人心的、精妙的抽象概念，这里的建筑形式千差万别，但都定义了拉脱维亚建筑的独特性。

contemplative hideaways from urban rush even in the city.

Latvia: Architecture Unfolding traces the architecture scene in Riga and Latvia mostly during the last 25 years that distinctively mark the acceleration and condensation of fragmented ideas, processes and practices in culture, unveil the hunger to understand what it means not only to be ecstatic about the country's regained independence in 1990 from the falling USSR, but also to actually cultivate it. Essentially it has been a question what makes Latvia extraordinary, what, perhaps, is Latvian-ness in the larger context of Europe and which are the common features that we share with the countries of uninterrupted development or on the opposite with the other newborn formations. Approaches to defining uniqueness in architecture have varied vastly through the past 25 years – from folklorized reductions of our fragmented history to the sleek and refined abstractions that arrives at shapes and programs that lift our hearts.

*Note: The original article was written in 2016, the 25th year of Latvia's independence.

* 编注：本文原文撰写于 2016 年。2016 年是拉脱维亚国家独立第 25 年。

Overview of Latvia

拉脱维亚概览

This book is divided into four areas: Old Town of Riga, Central Riga, Left Bank of the Daugava River, and Non-urban Environments. Riga has kept the medieval urban fabric. "Old Town" in our book means the area which used to be surrounded by the city wall and its neighborhood. In the early of 19th century the area outside the wall was developed rapidly and now has the main core of Riga City, so we call this area outside the wall "Central Riga". The following shows the divisions.

本书内容可分为四个地区:里加老城,里加新城,道加瓦河左岸,非城市环境。里加保留着中世纪的都市肌理,本书所说的"老城"代表过去被城墙围住的区域;"里加新城"则是指19世纪初发展迅速的城墙外的区域,今天里加市的核心也在这里。右图展示了分区范围。

Central area of Riga (scale: 1/100,000)／里加中心区域(比例：1/100,000)

Legend／图例

Central Riga
里加新城

Old Town
里加老城

Left Bank of the Daugava River
道加瓦河左岸

Non-urban Environments
非城市环境

The Historic Centre of Riga (HCR)
里加市历史中心区（$4.38\ km^2$）

HCR Protection Zone
*HCR*历史保护区（$15.74\ km^2$）

Contemporary works
当代建筑师作品

00

Historical Buildings
历史建筑物

D

Non-urban environments area
非城市环境区域示意图

History:

1198	Beginning of the crusades to Christianize the Baltics
1201	Foundation of the City of Riga
1282	Riga becomes an important trading post and joins the Hanseatic League
1583	The Livonian Confederation ceases to exist and the rest of Latvia comes under direct control of Poland-Lithuania
1586	The Roman Catholic and Lutheran catechisms printed as first books in the Latvian language
1621	Riga is conquered by Sweden under King Gustavus Adolphus and Eastern Latvia remains under the Polish rule
1710	Riga surrenders to the forces of Russian Tsar Peter the Great and Vidzeme comes under Russian rule
1772	Eastern Latvia becomes part of the Russian province of Polotsk
19 C	National Awakening in Latvia and Industrial Revolution
1918	The Republic of Latvia is proclaimed
1920	The Latvian–Soviet Peace Treaty was signed
1940	Latvia merged with the Soviet Union
1990	Declaration On the Restoration of Independence of the Republic of Latvia
1991	The Soviet Union recognizes Latvian independence and Latvia becomes a member of the United Nations
1997	World Heritage Committee inscribes The Historic Centre of Riga in the UNESCO World Heritage List
2004	Latvia becomes a member of NATO and of European Union
2014	Latvia becomes 18th state to join the Eurozone

Source: http://www.latvia.eu/geschichte/history-latvia-timeline

历史：

*1198*年	北方十字军入侵波罗的海地区，使其基督教化
*1201*年	里加建城
*1282*年	里加成为重要的贸易站，加入了汉萨同盟
*1583*年	利沃尼亚邦联不复存在，拉脱维亚其他地区由波兰立陶宛联邦直接控制
*1586*年	罗马天主教和路德教的教义被首次印刷成拉脱维亚语版
*1621*年	里加被瑞典国王古斯塔夫二世·阿道夫征服，但东拉脱维亚仍由波兰统治
*1710*年	里加向俄国沙皇彼得大帝的军队投降，拉脱维亚中部的维泽梅归俄国统治
*1772*年	东拉脱维亚成为当时俄国波洛茨克省的一部分
*19*世纪	拉脱维亚民族觉醒，开启工业革命
*1918*年	拉脱维亚共和国宣布成立
*1920*年	签订《拉脱维亚-苏俄和平条约》（里加条约）
*1940*年	拉脱维亚与苏联合并
*1990*年	通过《关于恢复拉脱维亚共和国独立主权的宣言》
*1991*年	苏联承认拉脱维亚独立，拉脱维亚成为联合国成员国
*1997*年	世界遗产委员会将里加历史中心列入联合国教科文组织世界遗产名录
*2004*年	拉脱维亚成为北约和欧盟的成员国
*2014*年	拉脱维亚成为第*18*个加入欧元区的国家

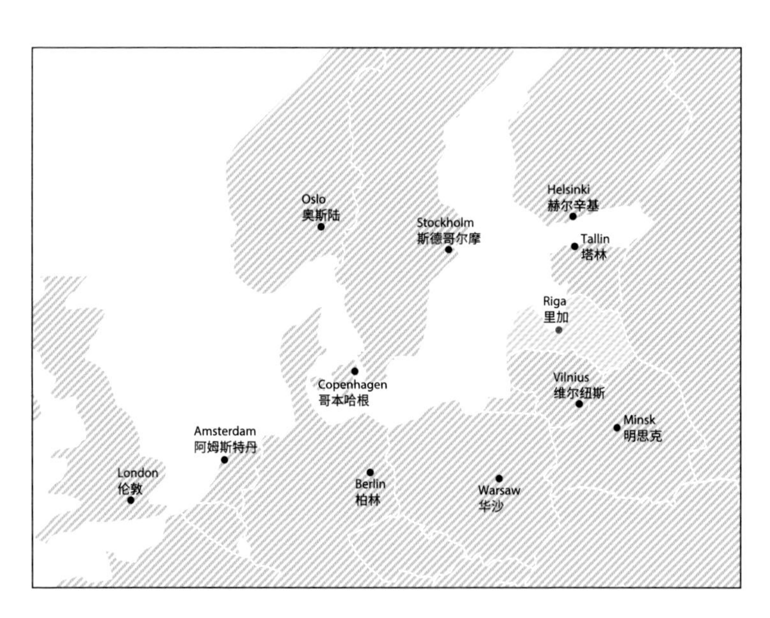

Northwestern Europe／欧洲西北部

General information

Latvia

Total area	64,573 km^2
Length of land boundary	1,368 km
Length of sea border	498 km
Population (2021)	1,893,223
Population decrease rate (2020)	-0.64%
Capital	Riga
Official language	Latvian
Ethnic groups (2021)	62.7% Latvians, 24.5% Russians, 3% Belarusians, 2.2% Ukrainians, 7.6% Others
Administrative divisions of Latvia	110 municipalities with subdivision interritorial units of parishes and towns, 9 republican cities

Source: http://www.csb.gov.lv

基本信息：

拉脱维亚

总面积	*64,573*平方千米
陆地边界线长度	*1,368*千米
海上边界线长度	*498*千米
人口（*2021*）	*1,893,223*
人口自然增长率（*2020*）	*-0.64%*
首都	里加
官方语言	拉脱维亚语
人口构成（*2021*）	*62.7%*拉脱维亚人 *24.5%*俄罗斯人 *3%*白俄罗斯人 *2.2%*乌克兰人 *7.6%*其他
拉脱维亚行政区划	按教区和城镇的领土单位细分的*110*个自治区和*9*个直辖市

Riga

Total area	307 km^2
Water	48 km^2 (15.8%)
Riga Planning Region area	10,133 km^2
Population (2021)	620,974 (32.7% of Latvia's population)
Number of the destrict	6

Source: http://www.latvia.eu

里加

总面积	*307*平方千米
水域面积	*48* 平方千米（*15.8%*）
里加计划区面积	*10,133*平方千米
人口（*2021*）	*620,974*（拉脱维亚人口的*32.7%*）
行政区数	*6*个

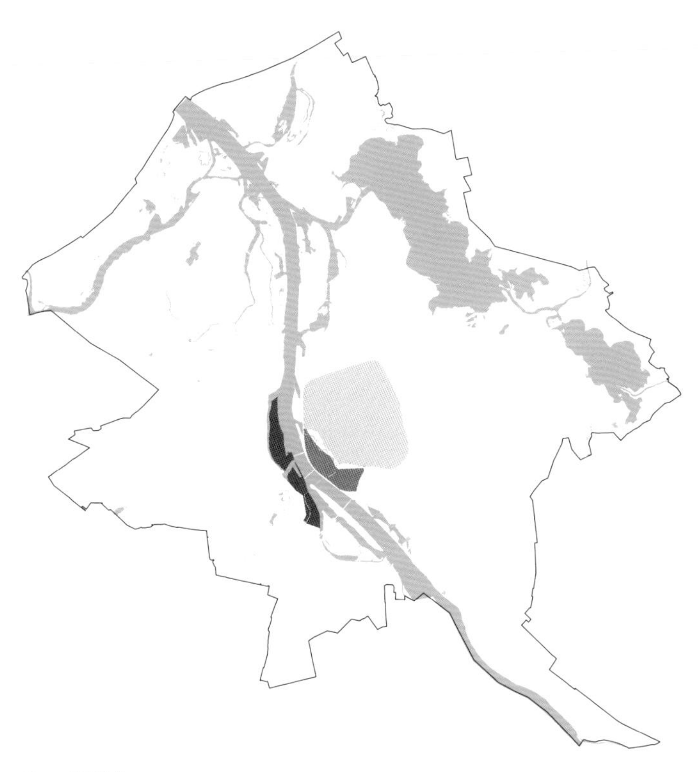

Riga／里加市

Essay:

The speckled capital of a monocentric country

Vents Vīnbergs

论文：

首都一级集中 *1 国家的斑斓首都

本茨 · 宾伯格斯

A peculiar pattern is frequently overlooked whenever one talks about forces that shape the cityscape of Riga. It is the successive marriages of convenience between the age-old city and its alternating rulers. The current union with the young state of Latvia is a new experience for both parts. Regardless of the upcoming official centenary of the state (but due to its history of abuse and the recent opportunity to start things over), Latvia is still in its troubled adolescence. Whereas Riga, being more than eight times older than its current spouse, bears a vast experience of dealing with various possessors and foreign influences. Age disparity matters, since the well-established spatial features that constitute Riga's identity and contribute to its acclaim are now challenged by the shady personal mythology of the coming-of-age country. A closer look at this bilateral jostle and its architectural manifestations may prove the arranged marriage metaphor not to be that whimsical at all.

The renowned flair that oozes in central Riga up until today (particularly, within the roughly 450 hectares of the listed UNESCO World Heritage site) was laid out in the master plan of 1856, designed by municipal architect Johann Daniel Felsko and his 23-year-old assistant Otto Dietze (fig.1). Similar large-scale rebuilding of cities was taking place all over Europe, but in few places was modernization executed with such a care for continuously accumulated historical traits. In Riga, as well as in Vienna at exactly the same time, the medieval inner core was kept distinct and almost intact, but the vast space newly freed of fortifications around the old town was turned into a park belt with an old moat, now the City Canal, winding through and delineating the imprint of former bastions and ravelins. The already formed outer suburbs with their rectangular street pattern now became the center itself, encircled by the newly introduced railroad. Alongside that, industrial areas developed, adding significantly to the wealth of Riga, whose main source of income traditionally was the maritime trade. Thus the city layout itself can be read as a textbook of European urban planning history in situ. This also applies to the Left Bank suburbs across the river, that were never deliberately planned up until the 20th century, and whose unprompted sprawl is no less intriguing.

Within the next 50 years the population of Riga increased ten times, reaching half a million just before WWI[1]. The year 1913 arguably was the peak year of the city's prosperity and exuberance, even if compared with the present day. Thus the reference point, by which success and failures of Riga are measured, is the booming years of the early 20th century (fig.2). In the same terms in which Walter Benjamin[2] describes the ever present feel of Baron Haussmann's Paris, and by which Rome is regarded as the baroque city. As always proud and ambitious itself, Riga still remained just a provincial trading outpost in the eye of its conquerors up until mid-19th century. Since then – a self-confident, yet still provincial metropolis within the firm setting of Russian empire.

The predominant bourgeois milieu of central Riga is recognizable, although a few streets there are homogeneous in terms of style or period. Even more – one may find there a wide variety of stylistic expressions of the same period architecture. It is the result of ever present assertiveness and

每当人们谈论起里加城市景观的演变时，往往会忽略一个特殊的模式，即这座古都与其不断更替的统治者之间的政治联姻。如今，里加又与年轻的拉脱维亚共和国结为连理，这对双方而言都是一次全新的体验。从拉脱维亚的角度来看，尽管即将迎来正式成立一百周年的纪念日（这是拉脱维亚历经屈辱后重新出发的契机），这个国家仍然处在坎坷的青春期。里加则比它现在的伴侣年长了8倍有余，在与统治者打交道和应对外来影响上有着丰富的经验。二者的年龄差导致了一个问题：即使这个国家进入成年期，有关它的传言仍会给里加特有的城市空间带来冲击。但正是这样的空间构成了里加的个性，提升了她的声誉。仔细观察这种双边角力及其在建筑上的表现，你会发现用包办婚姻来形容这种状态，并不离谱。

里加中心（特别是被列入联合国教科文组织世界遗产名录约450公顷的历史街区）今天仍然流露着的高贵气质，这可以追溯到1856年由市政建筑师约翰·丹尼尔·费尔斯科及其23岁的助手奥托·迭泽做的总规划图（图1）。尽管整个欧洲都有类似的大规模城市重建，但很少有地方在进行现代化建设时还能如此关怀历史积累。与同时期的维也纳一样，里加中世纪的内核得到了几近完整地保存。老城周围的防御工事被拆除后，这片广阔的空间变成了环城公园绿化带，老护城河（也就是今天的城市运河）蜿蜒而过，勾勒出昔日棱堡和半月堡的印记。网格化的郊区变身为城市中心，新铺设的铁路将其环绕。另外，随着工业的发展，过去主要依赖海上贸易的里加获得了更多财富。可以说，里加是欧洲城市原址规划的活范本。道加瓦河左岸的郊区也是一个耐人寻味的案例，那里直到20世纪之前都未被精心规划，随后却自发扩张为了城市郊区。

其后50年里，里加人口增加了十倍，在“一战”前就达到了50万[1]。繁荣兴盛的景象在1913年迎来巅峰，即使今天也不可同日而语。因此，20世纪早期突飞猛进的那些年就成了衡量里加城市发展进退的标准（图2）。当时的辉煌程度，可媲美被瓦尔特·本雅明[2]形容为“永恒之都”的、由奥斯曼男爵改造的巴黎，以及被视为巴洛克城市的罗马。然而，尽管里加向来自信满满、雄心勃勃，但19世纪中期前她仅仅是统治者眼中一个省级贸易中转站；被强大的俄国吞并后，也不过是一座省级大都市。

布尔乔亚*2氛围在里加市中心格外浓郁，即使有些街

3

unspoken rivalry within the community itself. Riga first and foremost is a civic undertaking. The city had not been a seat of any state power for the most part of its history, thus there are very few examples of imposed architecture, representing the governmental authority or ideology, except for a few references to the ruling power in public architecture and street names.[3] Three major oversized churches of the Old Town, for example, are not a kind of past ecclesiastical showiness, but an outcome of continuous political competition between different citizen factions. The obvious proof of that is the scarcity of monuments of conquerors, rulers or even prominent citizens of Riga, urban feature so typical elsewhere in Europe. The major communal endeavor of Rigans was the representation of their own virtues and respectability. The same can be said about the speckled appearance of the streetscape, although it is always conformed to practical building regulations of the day. For example, the celebrated Art Nouveau of Riga is so diverse not only because of various decoration manners, but also because of different attitudes that it ought to represent. The pre-1914 architecture of German patricians is full of Wagnerian sentiments, whereas Russian merchants and wealthy Jews were appropriating either traits of imperial capital or the flamboyant pageantry of Mikhael Eišenstein. The emerging community of prosperous Latvians often blended in or patronized the rising national romanticism in their commissions. The shared passion of all were expressive turrets on street corners, turning the tightly packed city blocks into fancy middle-class fortresses (fig.3). The unique oddity is the high number of small-scale wooden buildings, interrupting the otherwise seamless five-story streetscape. Paradoxically, the survival of these humble and flammable structures of pre-industrial Riga is a consequence of two wars, economic recessions and political turmoil, that slowed down their gradual replacement by the contemporary architecture of a current period. The appreciation of their historical and architectural significance, however, is a very recent trend.

The long-standing civic self-awareness of Riga was contested for the first time in 1918 when it became the national capital of newly proclaimed Latvia. And it has been so ever since. The new status required different self-image. That inconveniently coincided with the same task of the new state. Post-war hardship, however, delayed potential disagreements between new partners. New modernist ideas, which were quickly adopted by Riga and shaped into airy local functionalism, were suitable for self-reinvention of both. State institutions moved into the existing pre-war buildings, and government presence in Riga remained conveniently inconspicuous up until the pro-fascist coup of 1934. After seizing power the nationalist dictator initiated an imposing project of reshaping Riga in order to rid it of "self-indulgent architecture of the former oppressors". Large areas of the medieval town were cleared to give space for new government buildings, pseudo-classical, somber and intimidating in their appearance (fig.4). The recent functionalism seemed too cosmopolitan to represent "the ultimately unified nation". Few of his undertakings, however, were completed or even started when the WWII broke out and Latvia lost its independence for the next 50 years. Almost one third of the old town alone was destroyed due to interrupted rebuilding plans and wartime bombing. But rather than a physical devastation the biggest blow to the city, perhaps, was the loss of its multicultural diversity.

During the Soviet occupation, although rather dilapidated and with huge ripped chunks, the historical substance of Riga remained inevitably present. There were again radical and unrepentant plans to rebuild central Riga, but the government never found sufficient funding for that, and an everyday practicality took place. The fractured

道的风格符合建造年代，但更多是同一时期的建筑有着五花八门的表现形式。这要归结于本地居民一贯的强势和无声的对抗。里加首先是一项公民事业，这里历史上的大部分时间都不是国家政权所在地，因此象征政府权力和意识形态的宏伟建筑并不多见，仅有一些公共建筑和街道命名迎合了当时的统治阶层[3]。例如，老城中三座巨大的教堂，就不是出于一贯的教派炫耀，而是不同市民派别政治竞争的结果。还有一个明显的证据——在欧洲其他城市中相当典型的为征服者、统治者或伟人而建的纪念碑，在这里极其罕见。里加人在公共事业方面，更多是致力于展现他们的美德与尊严。同样的倾向也反映在了五彩斑斓的街景上，尽管建筑都遵循了当时的建筑规范。举例来说，里加著名的新艺术风格[*3]之所以富于变化，不仅在于装饰手法的多样，还因为它们需要展示出彼此各异的态度。1914年之前，德国贵族的宅邸充满了浪漫主义作曲家瓦格纳的情调，而俄罗斯商人和富有的犹太人则欣赏帝都风格和米哈伊尔·爱森斯坦[*4]富丽堂皇的设计。与此同时，逐渐走向繁荣的拉脱维亚人把日渐高涨的民族浪漫主义融入建筑。所有人都热衷于在街角立起富有表现力的塔楼，将拥挤的城市街区变成花哨的中产阶级城堡（图3）。然而奇特的是，在成片的五层楼中，竟有着不计其数的小木屋。这些简陋、易燃的建筑是里加工业化之前的遗存，两次世界大战、经济萧条和政治动荡减缓了当代建筑取而代之的步伐。直到最近，它们的历史意义和建筑意义才逐渐被认可。

1918年拉脱维亚宣告独立，并将新首都定在里加时，根深蒂固的公民意识第一次遭遇了挑战，从此摇摆不定。新的首都地位需要里加改头换面，这与一个新的国家的任务不谋而合。然而，在“战后”的困境中，这对新人甚至无暇面对彼此之间的矛盾。尽管新的现代主义理念对二者的自我革新都很适用，但只有里加迅速接受并独自走上了轻快功能主义道路。国家机关则搬进了战前建筑中，始终低调行事。1934年，一支亲法西斯组织发动了政变。掌权后信奉民族主义的独裁者卡尔利斯·乌尔马尼斯启动了一个重整里加的重大项目，以清理“前任压迫者引以为傲的建筑”。中世纪老城的大片土地让位给了新的政府大楼，它们仿古典主义的外观看起来冷硬阴森，令人生畏（图4）。在他看来，新潮的功能主义似乎过于国际化，无法代表这个“终极统一的国家”。然而他的举措鲜少实现，甚至大部分尚未开始，“二战”就已爆发，拉脱维亚丧失了独立主权，这种状态又

持续了50年。重建计划的中断，战时的轰炸，导致老城约三分之一被毁。但对这座城市而言，最大的打击或许不是物质上的破坏，而是文化多样性的消失。

苏联时期的里加残破不堪、千疮百孔，但城市仍有历史遗存。重建里加中心的计划又被提出，方式极端且毫无反省之意。由于政府始终没有筹够资金，从而改为了更实用的方案。支离破碎的城市肌理被现代主义建筑和基础设施填补、修复，有些还做得相当精良。当地民众发起了古迹保护运动，虽然气势不足，但至少算作一股阻力。此时老城已经被视为一种文化资产，因此维护街景，大部分重要建筑被修复，还增加了少量仿古建筑。

独立以来，今天的里加再次成为这个独立国家在政治、经济和才智方面的引擎，全国一半的人口都集中在首都地区。但这种相互依存的关系往往会带来不便，全球化也无法缓解。为了重新进入自由市场，这座城市快速经历着战后西方城市发展的每一个阶段（完整或部分）：快速郊区化、汽车导向型发展、房地产泡沫及破灭、引发微小火花的反文化运动、士绅化……在投资方的压力下，许多

urban fabric was steadily repaired with modernist infills and infrastructure objects, occasionally rather fine ones. And a monument protection movement arose among locals as a rather timid, but tolerated form of opposition. The Old Town was listed as a cultural value already under the Soviet rule, thus most significant buildings were restored and few replicas were added to cure the streetscape.

Now, for 25 years Riga is again political, economic and intellectual powerhouse of the independent state. Its metropolitan area is home to half of the country's population. But this interdependence is often inconvenient, and globalization does not make it easier. Re-entering free market conditions forced the city to catch up in acceleration with all successive stages in the post-war history of Western urban development, both faulty and sound: rapid suburbanization, submission to car-oriented development, a real estate bubble and its bursting, tiny sparkles of counterculture, gentrification processes, etc. Under the pressure of investors many impetuous decisions are made and most of new buildings are generic and bland, justified by slogan "It's just business." New forms of use and renovation are required for the vast amount of architectural heritage that does not represent its original purpose any more. And there are already two generations of local architects, who specialize particularly in reinvention of this heritage. Their work and expertise has recently become the most rewarded among professionals. Thestate, in the other hand, is clinging to popular conservatism in order to prove its continuity, thus falling prone also to pre-war nationalist prejudices[1]. Anything historical usually fits traditionalist mythology indiscriminately, and Riga, being the oldest continuous entity in the country and the capital, is confined in it. The result of this is various forms of pastiche and replicas, often even alien to history and context of certain localities. The general sense, however, is, that this adolescent confusion can be overcome. There is a long-standing tradition of the visual depiction of Riga, dating back to late medieval period. It is the Old Town's Hanseatic silhouette, that has been turned into a popular image of two-dimensional cardiogram. And it pulsates vividly.

Note.

1. By 1840 the population of Riga was – 60,000, one thirds of which – in the densely packed Old Town. In 1913 the population was 517,500, and only 35% of inhabitants were born there.
2. Walter Benjamin, Paris, Capital of the Nineteenth Century, 1939
3. For more than 100 years (1818-1923) the main axial street of the city was named after Russian Emperor Alexander I. During the soviet occupation it was Lenin's Street. Now, as well as in pre-WWII years it is Brivibas (Freedom) Street.And it took more than a decade of disagreement between city and Latvian state to find a site for the Freedom Monument (1935), now located in the same spot, where once a statue of Peter the Great stood for a very short period.
4. In 1973 the reconstruction of St. Peter's Church tower was completed, 32 years after the destruction of the 18th century original in 1941, then the tallest wooden structure in Europe (123 m). In 70s and 80s Polish restorators were invited to renovate several pre-war streetscapes in the Old Town, in appreciation of their expertise in recreation of destroyed Warshaw.

Fig.1: "Riga Mater plan" (1856, detail) by Johann Daniel Felsko and Otto Dietze. New boulevard belt around the Old Town is drawn. From the collection of Museum of Riga

Fig.2: A postcard to show Albert Street in Riga before WWI., published in Berlin. From the collection of the Digital Forum for Central and Eastern Europe.

Fig.3: It is Brivibas (Freedom) Street with Art Nouveau apartment building from 1910s (on the right) and Functionalist apartment block from 1930s.

Fig.4: View of Iirgu Horse street in Vecrigat, showing the Ministry of Finance, 1938 by architect Aleksandrs Klinklavs on the left. This is what the most of the Old Town would have looked if all the rebuilding projects by dictator Ulmanis were executed.

Vents Vīnbergs (1979–) is a graduate of the Riga Technical University, becoming an architect in 2004. Since 2007, however, he participates in the profession by writing about it. He is the author of over a hundred publications in the media of Latvia and its neighbouring countries as well as the book series Process (2008–2013) and the book *V*X*. Latvian Architecture Since 1991 (2011). He has a regular column in the largest daily newspaper Diena, has been a set designer in theatre and co-curator of exhibitions. V nbergs is interested in the role of the individual in architectural processes.

而已"敷衍了事。大量建筑遗产最初的功能不再适用，新的用途和翻新有待实现。如今已有两代专门从事建筑遗产活化的本地建筑师，他们能够凭借专业的工作和知识，拿到业内最高的报酬。另一方面，国家仍固守着普遍的保守主义，试图证明它的连续性，因而也容易落入战前的民族主义偏见[4]。通常，任何古老的事物都能被生搬硬套上一个相应的民间神话，里加作为这个国家的首都和历史最悠久且未间断的存在，自然无法幸免。最终的结果，便是出现各种形式的拼凑和复制，而且往往与所在地的历史、背景格格不入。但总体而言，这种"青春期"的困惑都可以解决。另外，里加自从中世纪晚期开始就有描绘城市的传统，老城在汉萨同盟时期，其建筑轮廓就已经变成一幅颇受欢迎的心电图风格。直到今天，这颗心脏仍有力地脉动着。

注释：

1. 1840年里加拥有6万人口，其中三分之一密集生活在老城。1913年的人口达到51.75万，仅有35%为土生土长的居民。
2. 瓦尔特·本雅明，《巴黎，19世纪的首都》，1939
3. 一百多年来（1818-1923），这座城市的中轴大街始终以俄国皇帝亚历山大一世的名字命名。在苏联时期，它改为了"列宁大道"。而现在以及"二战"之前，它是"布里维巴斯大道"（意为"自由大道"）。为了确定自由纪念碑（1935）的选址，里加市和拉脱维亚的管理者还经历了十年多的讨论，最终一致同意，设在一个曾短暂矗立过一座彼得大帝雕像的地方。
4. 1973年，圣彼得教堂塔楼（始于18世纪）的重建工程竣工，距离1941年被毁已经过去了32年。而最初的塔楼是当时欧洲最高的木结构建筑（123米）。20世纪七八十年代，为整修老城几处战前的旧街道，还特地邀请了波兰修复师，他们经验丰富，曾复原过饱受摧残的华沙。

图1："里加总规划图"（1856，细节图），约翰·丹尼尔·费尔斯科和奥托·迭泽绘制。可看到环绕老城的新大道。来自里加博物馆的收藏。

图2：一张明信片，画面中是"一战"前里加的阿尔伯特街，柏林印制。来自中东欧数字论坛的收藏。

图3：布里维巴斯（自由）大道，以及20世纪10年代的新艺术风格公寓楼（右侧）和20世纪30年代的功能主义公寓楼。

图4：里加老城的伊尔古马街，左侧是建筑师亚历山大·克林克拉夫斯在1938年设计的财政部大楼。如果独裁者卡尔利斯·乌尔马尼斯所有重建项目都实现的话，这就会是老城大部分的面貌。

*1译注：首都一级集中是指，一个国家的政治、经济、文化、人口以及社会资源和活动过度集中于首都及其周边区域的问题。如日本的"东京一级集中"现象。此处也可译作"里加一级集中"。

*2译注：布尔乔亚（Burgensis），一般指资产阶级，小布尔乔亚，即小资产阶级。

*3译注：新艺术风格（Art Nouveau），是指19世纪末20世纪初流行于欧洲和美国的装饰艺术和建筑风格，以装饰繁复华美的外立面、充满流动感的线条造型为主要特征。

*4译注：米哈伊尔·奥西波维奇·爱森斯坦（Mikhael Osipovich Eisenstein，1867-1921），里加当时最著名的本土建筑

Latvian Landscapes

拉脱维亚风光

pp. 26–27: Bird's eye view towards the northwest from St. Peter's Church in the Historic Center of Riga. Looking at Riga Cathedral in front, Riga Castle at the rear, and the left bank of the Daugava river. All photos on pp. 26–255 except as noted by Shinkenchiku-sha.

第26-27页：从里加历史中心的圣彼得教堂向西北俯瞰，由近至远依次是里加大教堂、里加城堡和道加瓦河左岸。

pp. 28–29: Evening view of the Historic Center of Riga. The skyline with towers, TV and Radio tower, and the Vanšu Bridge are seen from the left bank of the Daugava river.

第28-29页：黄昏下的里加历史中心。从道加瓦河左岸远眺，可以看到众多塔楼、广播电视塔以及凡苏大桥构成的城市天际线。

pp. 30–31: Aerial view over island of Ķīpsala on the left bank of the Daugava river. The island, gradually built up with the small wooden residential buildings seen in front, starting from the 17th century, is now developing into a luxurious residential district. In the back right, high-rise buildings are under construction.

第30-31页：鸟瞰道加瓦河左岸的基普萨拉岛。近前的木构建筑群是从17世纪开始建造起来的渔业小木屋，现在正发展为高级住宅区。远处可见正在建设中的高层建筑群。

pp. 32–33: Typical landscape of rural areas in Latvia, located near Ulmale at the coastline of the Baltic Sea in western Latvia. Traditionally a collection of buildings surrounded by trees constitutes one farmstead, the next farmstead being some distance away.

第32-33页：拉脱维亚乡村的代表风景。这里地处拉脱维亚西部，靠近波罗的海沿岸的乌尔马累。传统上，被树木环绕起来的一个建筑群就是一座农庄，农庄之间通常相隔甚远。

Central Riga

里加新城

Elimination of the fortifications of Riga and growing prosperity of the city in the second half of the 19th century fostered rapid development of a new urban center with well-defined spatial and programmatic design principles for the newly acquired territory. Consequently, most of the buildings outside the city wall were built over a surprisingly short span of time at the very beginning of the 20th century. The formal architectural expressions range from Neostyles to first experiments with new forms of Art Nouveau and sensible attempts to define local forms of Art Nouveau, and to later pragmatic approaches of Art Deco and Functionalism, as well as modernism. Although buildings exhibit extraordinary varied character, the streetscape emanates a very balanced atmosphere. The exact moment of subtle transition, transformation, turning from one thing into another can be traced also within the new architecture in this area.

19世纪后半叶，里加的防御工事被拆除，城市日益繁荣。与此同时，新增城市用地有了明确定义的空间以及相应的规划设计，里加新城由此得以迅速发展。所以，里加城墙外的建筑多是在20世纪初一段极短的时间内被建造出来的。这里的建筑表现形式有对新风格到新艺术风格的初步探索，对新艺术风格本土化的谨慎尝试，后来又发展为对装饰艺术、功能主义以及现代主义的实践应用。尽管这些建筑表现出非凡的多样性，但整个街道的氛围非常协调。我们也能在这个区域的新建筑物中，感受到从一件事物过渡和转变到另一件事物的精确瞬间。

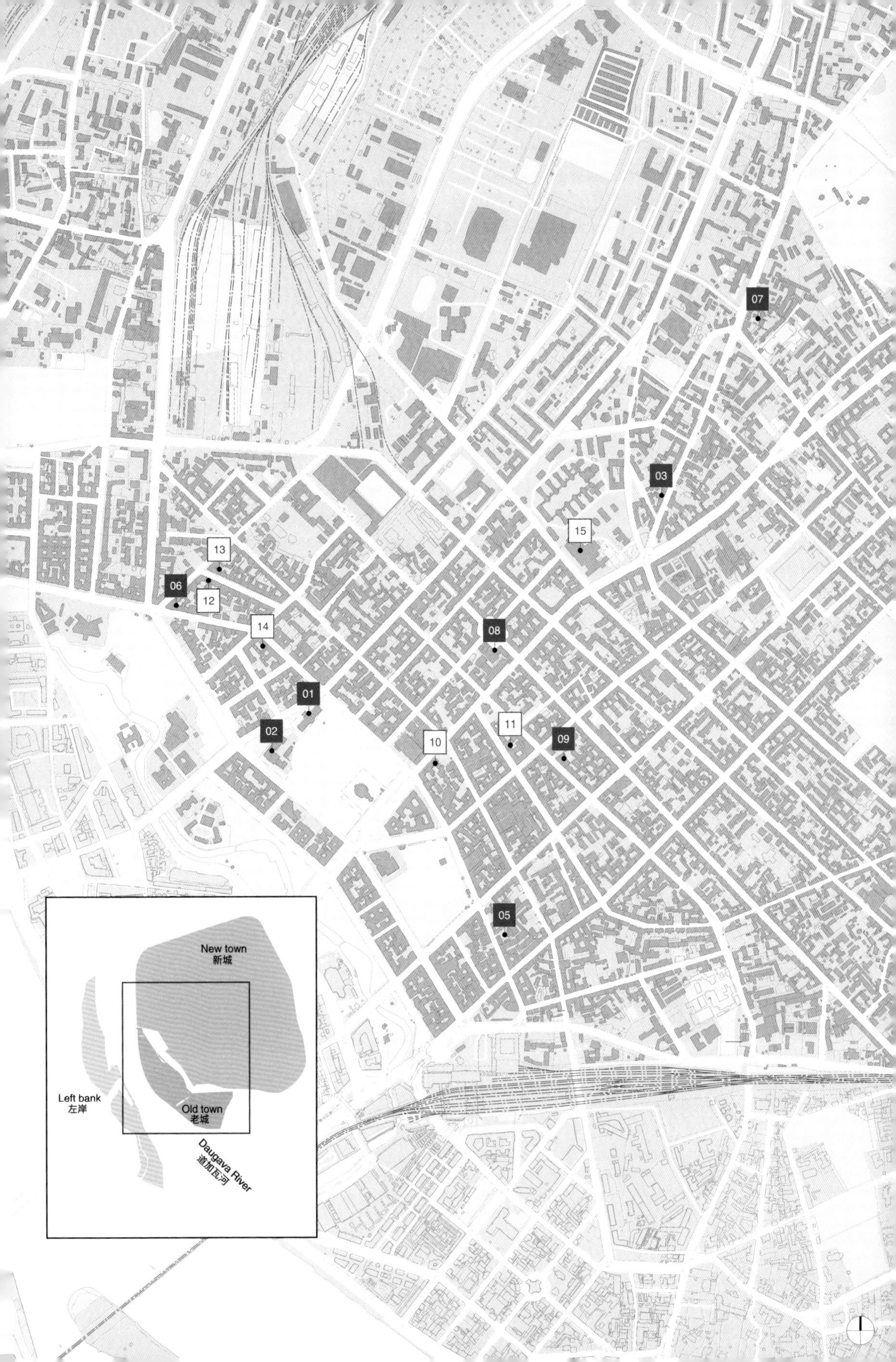
07
03
15
13
06
12
14
08
01
02
11
10
09
05
New town
新城
Left bank
左岸
Old town
老城
Daugava River
道加瓦河

Renovation and Extension of Latvian National Museum of Art
Processoffice and Andrius Skiezgelas Architecture
K. Valdemāra Str. 10, Riga
2016 (pp. 38–49)

Extension of the Latvian Academy of Art
SZK and Partners
Kalpaka Blvd. 13, Riga
2012 (pp. 50–55)

Mierīgi!
Fine Young Urbanists
Miera Str. 9
2014 (pp. 56–61)

Alekša Square
Fine Young Urbanists
Alekša and Tilta Str. crossing, Riga
2014 (pp. 62–63)

Renovation of the Bergs Bazaar
Zaiga Gaile Office
Block among Elizabetes, Marijas and Dzirnavu Str., Riga
1993- (pp. 64–73)

Staircase Renovation Art Deco
Sudraba Arhitektūra
Elizabetes Str. 21a, Riga
2013 (pp. 74–79)

House in the Courtyard
Sudraba Arhitektūra
Miera Str. 52a, Riga
2008 (pp. 80–85)

Office and Retail Premises Building in Riga, Baznīcas Str. 20/22
Arhis
Baznīcas Str. 20/22, Riga
2003 (pp. 86–89)

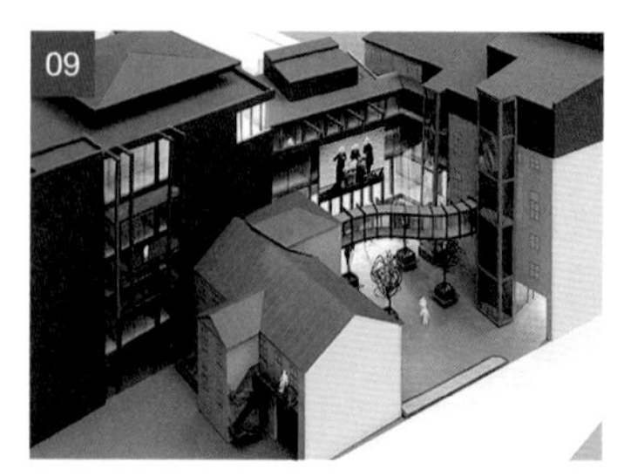

New Riga Theatre Reconstruction Project
Zaiga Gaile Office
Lāčplēša Str. 25, Riga
2014 (pp. 90–93)

Photo Credit
03: Photo by Kaspars Kursišs, courtesy of the architect /04: Photo by Ansis Starks, courtesy of the architect /06:Photo by Didzis Grodzs, courtesy of the architect

Apartments and Retail Premises
Aleksandrs Klinklāvs
Brīvības Str. 40
1934 (p. 95)

Atis Ķeniņš School
Konstantīns Pēkšēns, Eižens Laube
Tēsrbatas Str. 15/17
1905 (pp. 96–97)

Apartments
Eižens Laube
Alberta Str. 11
1908 (p. 98)

Apartments
(**Riga Art Nouveau Museum**)
Konstantīns Pēkšēns, Eižens Laube
Alberta Str. 12
1903 (p. 99)

Apartments and Retail Premises
Mihaels Eizenšteins
Elizabetes Str. 10a
1903 (p. 100)

Dailes Theatre
Marta Staņa, Imants Jekabsons, Haralds Kanders
Brīvības Str. 75
1959–1976 (pp. 101–103)

Processoffice and Andrius Skiezgelas Architecture

Renovation and Extension of Latvian National Museum of Art

K. Valdemāra Str. 10, Riga 2016

Process建筑师事务所，奥德雷斯 · 谢科茨格拉斯建筑师事务所

拉脱维亚国家艺术博物馆改造及扩建

里加，克里斯蒂安 · 瓦尔德马尔大街 10 号 2016

01

Credits and Data

Credits and Data

Project title: Renovation and Extension of Latvian National Museum of Art

Client: Riga City Council Property Department

Program of original building by Wilhelm Neumann: Restoration and new exhibition halls in the cupola and attic spaces. First floor replanned to include a conference hall, cafeteria, bookshop, children education rooms and museum administration offices.

Program of new extension: 3 exhibition halls, artwork storages, restoration workshops, supporting and technical premises

Location: 10A Krišjāņa Valdemāra St, Riga, Latvia

Dates: Competition 1st place 2010, technical project 2011–12, completion 2015

Architects: Processoffice / Vytautas Biekša, Rokas Kilčiauskas, Marius Kanevičius, Giedrius Špogis, Ježi Stankevič, Austė Kuliešiūtė, Miglė Nainytė, Giedrė Datenytė, Mantas Petraitis, Sandra Dumčiūtė, Povilas Marozas, Sandra Šlepikaitė Andrius Skiezgelas Architecture / Andrius Skiezgelas, Gilma Teodora Gylytė, Rasa Mizaraitė

Restoration: Arhitektoniskās Izpētes Grupa / Artūrs Lapiņš, Marina Mihailova, Guntars Jansons

Structural engineering: Engineers' office Būve un Forma / Jānis Prauliņš, Jānis Krasts, Kaspars Šņore, Olga Opolčenova, Solvita Šņore, Māris Grāvītis

Climate, water supply and sewer systems: Engineers' office Būve un Forma / Jeļena Uspenska, Tatjana Grava, Sergejs Pivovarovs, Rasa Vilka

Fire safety: US&L / Gendijs Kuzmins, Alina Kavalera; UG Projekts / Juris Šmits

Electrical and low voltage systems: Daina-EL / Ingus Mozaļevskis, Eduards Lūsis, Ilgvars Kozlovs, Viktors Grinčuks; Lafivents – Andris Krūmiņš, Nikolajs Bogdanovs; Telekom Serviss / Igors Musijenko

Lighting consultants: Think Light, Moduls

Landscape: Ainavu arhitekti / Daiga Veinberga, Līga Valdmane

Total area: 8,249 m²

Public space: 2,500 m²

Budget: 34 million euro

In 2010 the Riga City Council announced an international architectural competition for the reconstruction and extension of the Latvian National Museum of Art building. In 2010 the jury unanimously agreed that the best proposal was by the Processoffice.

The main building of the Latvian National Museum of Art was built between 1903 and 1905 for the needs of the Riga City Museum of Art and the Riga Society for Art Promotion. The main building of the museum is an architectural monument of national importance and has served its purpose without major repair for 105 years. It was therefore decided to reconstruct and extend the building with the aim of creating a contemporary museum infrastructure for the preservation and exhibition of artworks as well as to provide the optimal environment for people to be able to learn and spend their free time constructively in the museum space.

To solve the problem of a shortage of exhibition space, we offer the adjustment of existing unused building space and additional functions, as well as the formation of new exhibition spaces concentrated in the new extension, which is located below ground level. Placement of the new extension below ground level makes it possible to maximize the preservation of the building as a historical urban landmark. Next to the exceptional museum building, only a neutral concrete courtyard with an amphitheatre in brass is added.

Working on modernization of the museum building, we have concentrated on the development of the functional strategy, also drawing attention to redistribution of space and logical classification, grouping and connection of the existing and future museum premises and functions. In response to the needs of the museum, we have come to the following functional approach: Leaving the site free of any new ground-based volumes, we propose to construct the space for temporary exhibitions, storage with restoration workshops and ancillary facilities in the new extension below ground level. Above the underground extension, in between the park and museum building, we suggest a square – a stylized social space with an outdoor café, video projection capability, and art installations in the open air and other public events. The ground floor is adapted for the administrative and public functions of the museum, which can be accessed both from the street and park side. In addition to the existing exhibition space in the old building, we propose to use the cupola and part of the currently unused attic as exhibition spaces. For convenient and practical use we suggest redesigning part of the roof in a way that allows it to accommodate a terrace with picturesque panoramas, while admitting natural light into the second floor exhibition galleries through the original skylights.

pp. 38–39: Exterior view from the southeast. The museum stands in the Esplan de park. A new entrance of the southwest facade was built with public square. It has a 9 × 9 m glass covered atrium underground which visitors to the public square can look down into. pp. 40–41: View of the stairs towards underground from the main lobby. Photo by Norbert Tukaj, courtesy of the architect.

第 38-39 页：东南方向外景。美术馆位于艾斯普兰德公园，其西南立面建造有新的入口以及公共广场。此外，这里还建造有一个 9 米 ×9 米的地下中庭。中庭被玻璃所覆盖，路过广场的游客可俯瞰中庭。第 40-41 页：主厅通向地下的楼梯。

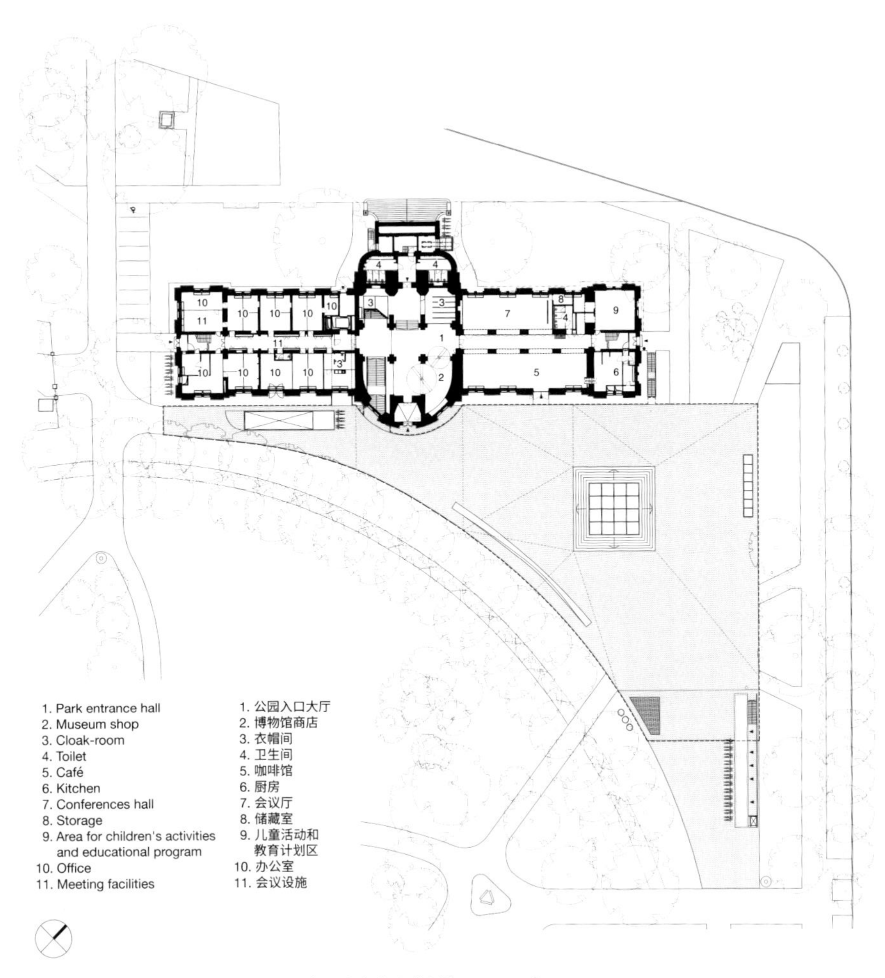

Ground floor plan (scale: 1/1,300) ／一层平面图（比例：1/1,300）

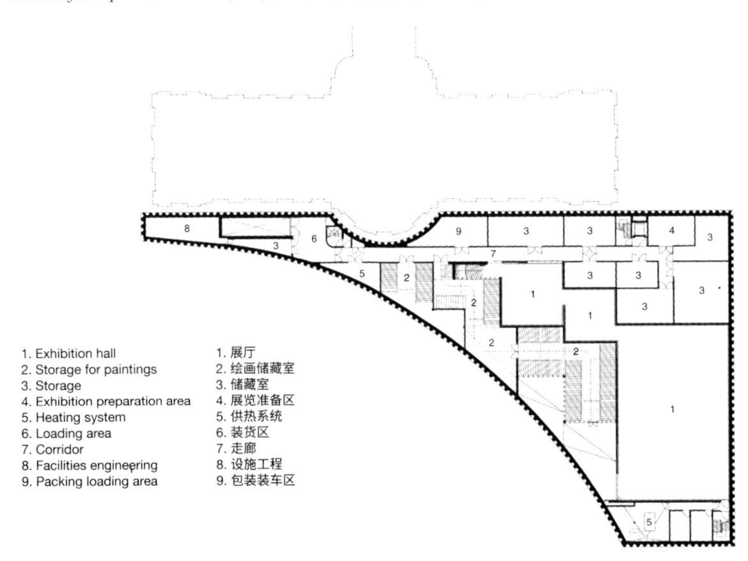

Basement floor plan ／地下一层平面图

2010 年，里加市议会为改造及扩建拉脱维亚国家艺术博物馆征集方案而举办了一场国际建筑设计竞赛。同年，经评委会一致认定我们提交的方案获得优胜。

根据里加市艺术博物馆和里加艺术推广协会的需要，里加于 1903 至 1905 年建造了拉脱维亚国家艺术博物馆主楼。这是一座国家级的纪念性建筑，并在其建成后加以使用的 105 年间，没有经历过重大整修。这次改造和扩建的目的在于打造一座用于保存和展示艺术品的当代博物馆，同时为人们提供学习、休闲的最佳环境。

为解决展览空间不足的问题，我们对已有闲置空间进行了调整，补充了附加功能；同时，还以地下层扩建部分为重心，创造新的展览空间。通过将新扩建的空间布置在地下，原有的城市地标性历史建筑得到最大程度的保留。此外，在这座独具特色的博物馆建筑一旁，我们仅建造了一个低调的混凝土庭院，院中有一座黄铜露天剧场，以凸显建筑本身的特色。

为实现博物馆建筑的现代化，我们专注在建筑功能策略的发展上，同时也关注空间的重新布局以及新旧建筑功能组织之间的逻辑分类、重组和相互连接。为满足博物馆的需求，我们采取了以下功能性方法。首先，不在地面上增建新体量，将临时展区、库房、修复工作室以及辅助设施均布置到地下的加建空间中。其次，我们建议在地下加建空间的上方，也就是公园与博物馆建筑之间的区域，建造一座广场。结合露天咖啡馆、露天影院、室外艺术装置等公共活动设施，这里可以成为一个独具风格的社交空间。另外，一楼为行政及公共功能而设计、改造，人们可以从街道或者公园进入其中。最后，除老建筑内已有的展览空间外，我们还将穹顶以及部分未经使用的阁楼改造为展厅。考虑到使用的便利性和现实性，我们提出重新设计屋顶，使其可以承载一个观景露台，并且保证自然光可以透过原有的天窗照亮二楼的展廊。

Opposite: Interior view of the underground atrium. It also functions as a skylight for the exhibition space. Photo by Sturmanis, courtesy of the architect.

对页：地下中庭内景。它还可以作为展览空间的天窗。

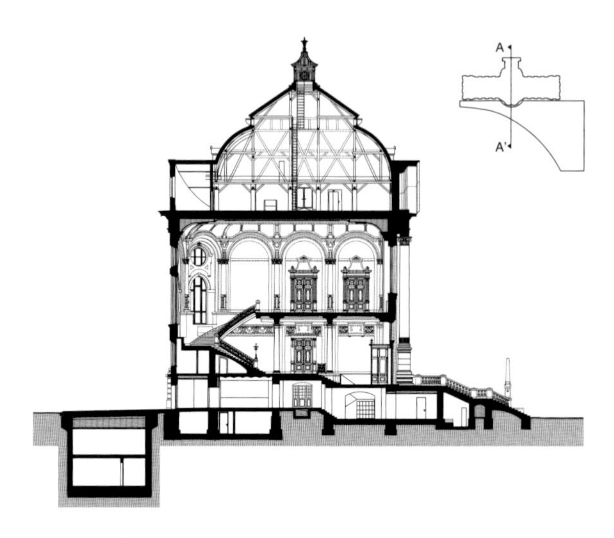

A-A' section (1/800)／A-A'剖面图（比例：1/800）

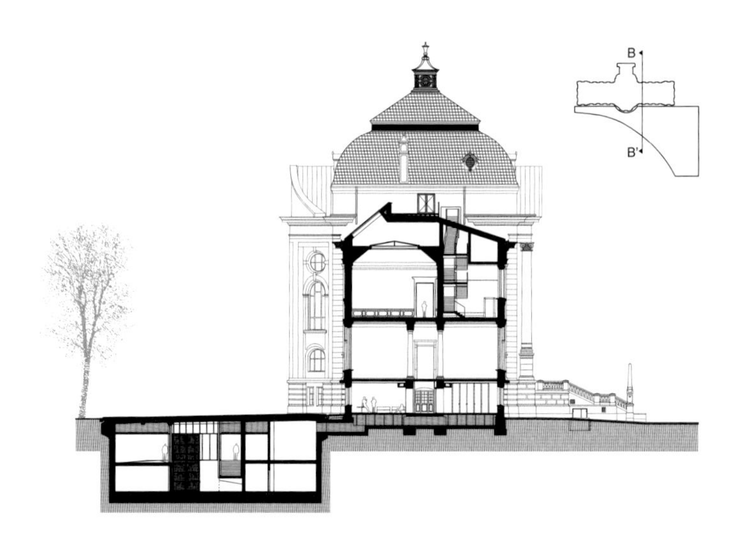

B-B' section／B-B'剖面图

pp. 46–47: View of exhibition space on the second floor. Opposite: View of exhibition space, formerly an unused attic, with natural light from skylights and renovated wooden roof constructions. All photos on pp. 46–49 by Norbert Tukaj, courtesy of the architect.

第46-47页：二楼展览空间内景。对页：展览空间内景。这里曾是一间未经使用的阁楼，有可以引入自然光的天窗和已被翻新的木构屋顶。

SZK and Partners

Extension of the Latvian Academy of Art

Kalpaka Blvd. 13, Riga 2012

SZK合伙人建筑师事务所
拉脱维亚艺术学院扩建
里加，卡帕卡大街13号 2012

The new exhibition and lecture hall of the Latvian Academy of Art is one of the first modern university extensions in many years and one in a series of others currently being built around Latvia with the help of EU funding. The extension was built on a narrow plot sandwiched between the old academy building (1905) and Esplanade Park in the historical center of Riga. Both sites are local monuments of architecture and part of the UNESCO protected World Heritage Site.

The new extension is a conversion of the old warehouse building in the academy courtyard. Built in 1948 it was stylistically matching but badly constructed post-war storage, yet it seemed important to incorporate three of the existing brick walls in the new structure both for aesthetic and practical reasons. The old walls with several new openings now provide muted daylight to the lecture rooms and together with the overhanging roof naturally protect the newly built glass walls from the direct sun.

Organized in two levels with an additional third level on the roof, the extension was lowered three meters deep to keep its low, backyard profile. To save costs and increase accessibility a covered slope was created instead of stairs and elevators. Use of the old brick walls allowed the new concrete structure and glass walls to be as simple as possible. Clarity in form and building materials was kept in the interiors that are left bare to show the plainness of concrete, glass and historical brick. The exhibition hall and rooms for lectures can be organized by mobile indoor walls, creating smaller studios and alcoves when necessary.

Credits and Data

Project title: Extension of the Latvian Academy of Art
Location: Riga, Latvia
Architect: SZK and Partners (Andis Silis, Guntis Ziņģis, Pēteris Kļava)
Dates: 2010–2012
Total floor area: 550 m² + 115 m² roof
Budget: 1,390,000 euros

拉脱维亚艺术学院的新展览与讲演厅是这里多年来第一批现代化大学扩建工程之一；也是欧盟资助下，拉脱维亚各地正在建设的项目之一。扩建建筑位于里加历史中心旧学院大楼（1905年）与滨海公园之间的狭长地块上。这栋学院大楼及滨海公园都是当地城市地标，并被联合国教科文组织列为世界遗产地。

新扩建的部分是对学院中庭内旧仓库的改造。这座仓库建于1948年，在风格上与周边的建筑相契合，却属于战后建造的劣质建筑物。尽管如此，出于审美及实用两方面的考虑，将其现有的三面砖墙整合到新建筑中似乎十分重要。旧砖墙上设置的几个新开口，为讲演室带来柔和的光线，同时悬挑的屋顶则形成了自然的保护屏障，以免新建成的玻璃墙受到阳光的直射。

扩建部分包含两层，而屋顶上的空间则作为第三层，为了塑造后院相对较低的轮廓，整座建筑的高度降低了3米。同时，我们使用一段有遮挡的斜坡来代替楼梯和电梯，既节约成本又提高了建筑的可达性。旧砖墙的使用令新的混凝土结构和玻璃墙尽可能得到化简。同时，建筑形式与材料的清晰明快在室内空间中也得以保持。混凝土、玻璃及裸露的旧砖块，彰显了建筑整体平整、朴素的特点。展览厅及讲演室的空间能够利用可移动内墙来组织，根据需要创建出较小的工作室或隔间。

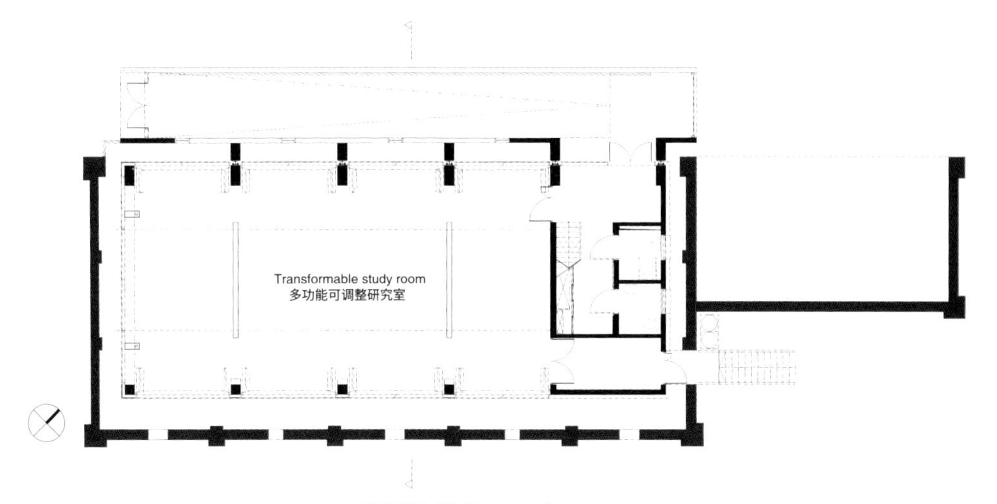

Ground floor plan (scale: 1/300)／一层平面图（比例：1/300）

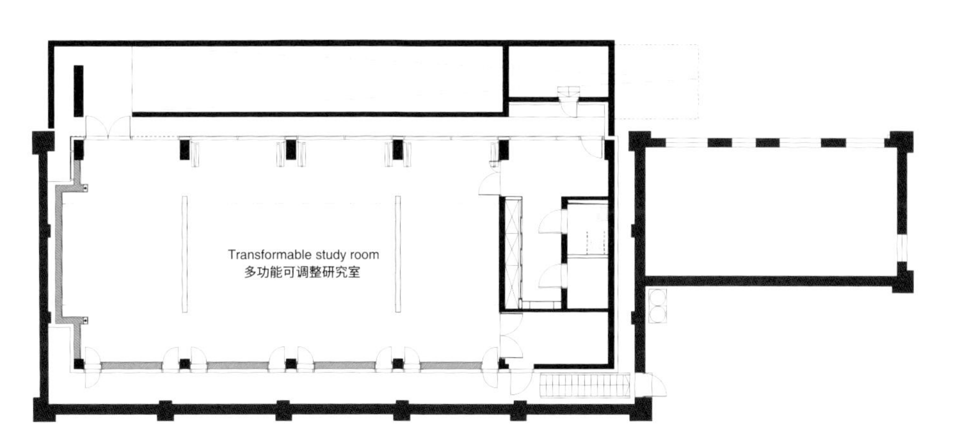

Basement floor plan ／地下层平面图

p. 51: General view from the northeast. Opposite: View of the two slopes. All photo on pp. 50–55 by Norbert Tukaj.

第 51 页：东北方向全景。对页：两条坡道。

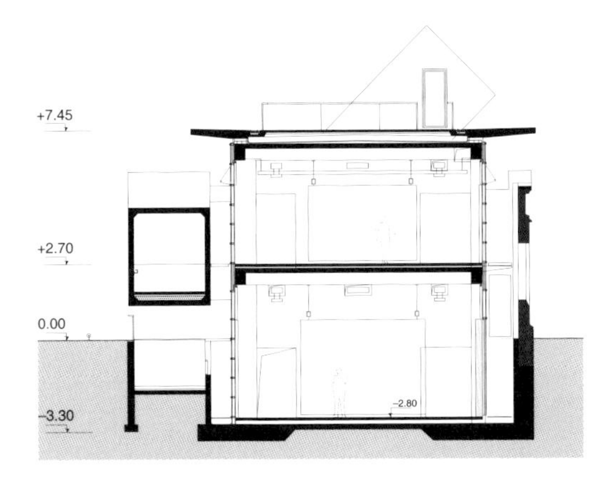

Section (scale: 1/300) ／剖面图（比例：1/300）

Opposite: View of interval between old wall and concrete structure. Preservation of the red brick finish was required because the new building is located in the monument protection zone. The brick wall works as blinds towards the south side, providing daylight thorough the windows, and allows the main volume to be inserted into inexpensive standard glass. This page: Interior view of the study room.

对页：旧砖墙与混凝土墙之间的空间。由于新建筑位于里加历史保护区，所以红砖必须被保留。而砖墙可以作为朝南的百叶窗，使阳光透过窗户进入室内；同时，它也使得在主体量中使用成本低廉的标准玻璃成为可能。本页：研究室内景。

Fine Young Urbanists

Mierīgi!

Miera Str. 9, Riga 2014

都市青年建筑事务所

米埃里基!

里加，米埃拉大街9号 2014

03

8 - 21
8 - 21
8 - 21
8 - 21
8 - 21
10 - 21
10 - 18
Wi Fi
9-22, 30-32

Miera Street in Riga is designed with tram and car traffic in mind, but its craft shops and cozy cafes attract an increasing number of cyclists and pedestrians, which often leads to a conflict on the rather narrow pavement. For the last four years we have been actively advocating a more humane approach to street and public space design in Riga. To prove that the street can be a space for both effective mobility and social life, we built a 14-m-long street section on a scale 1/1 with wider sidewalks and a bicycle lane in each direction. The mock-up was built in three days and remained in place for a week. We used this time to discuss street design with passers-by, local residents and businessmen, discovering an effective method of involving public in the design process – no one can pass through a vividly blue space without wondering about its purpose!
The meaning of the Latvian word mierīgi is "peacefully, easily".

Credits and Data
Project title: Mierīgi!
Client: Self–initiated / Miera Street Republic
Location: Riga, Latvia
Year: 2014
Architect: Fine Young Urbanists – Evelina Ozola, Toms Kokins
Partners: Riga 2014 European Capital of Culture, Survival Kit 6
Construction team: Guntis Jasotis, Elvijs Jamonts, Nauris Lā rmanis, Oto Ozols, Ivars Jankovskis, Edijs Vucēns, Evelina Ozola, Toms Kokins

里加的米埃拉大街最初是为有轨电车及汽车的通行而设计的，但随着街道两旁的杂货店和咖啡馆吸引了越来越多的骑行者和行人，这里狭窄的人行道经常会出现交通问题。在过去的四年里，我们积极主张用更人性化的方法来设计里加的街道和公共空间。为了证明这条街道可以成为有效的交通及社会生活空间，我们按照 1/1 的比例建造了一段 14 米长的道路，并在路两边各设有更宽的人行道和一条自行车道。整个模型在三天内完成，并在原街道上放置了一周。我们还利用这段时间与路人、当地居民以及这条街上的商人讨论街道设计，从中发现了一种可号召公众参与设计的有效方法——因为人们在走过这条鲜亮的蓝色道路时，就自然会去思考它的用途。

在拉脱维亚语中，“mierīgi”一词有“和平、轻松”之意。

Site plan (scale: 1/1,300)／总平面图（比例：1/1,300）

pp. 56–57: View of the west side. A bicycle lane installed by the project occupies part of the road. p. 58: View of the site from the northeast after the project was removed. Parking on the street is seen around Riga. p. 59: General view from the northeast. Opposite: Bird's eye view of the project. A car drives over the tram rails. All photo on pp. 56–61, expect on this page above, by Kaspars Kursišs.

第 56-57 页：西侧景观。项目设置的自行车道占用了部分道路。第 58 页：从东北方向看项目撤去后的用地。在道路上停车是里加的日常景象。第 59 页：从东北方向看到的全景。对页：项目鸟瞰。一辆汽车正在有轨电车的线路上行驶。

Fine Young Urbanists

Alekša Square

Alekša and Tilta Str. crossing, Riga 2014

都市青年建筑事务所

亚历山大广场

里加，亚历山大路和小杜塔路的十字路口 2014

04

Alekša Square is the first public square in Riga, co-designed with the local community in the 'problem neighbourhood' of Sarkandaugava. Following a special methodology, local residents generated ideas and discussed necessities for a new public space in several meetings; designers then translated their suggestions and necessities into a design for the square. The grid plan adds structure to the area and allows for unlimited variations for the elements. The square currently contains a ping-pong table, a board games area, a seating area, and a vertical gym – custom designed exercising equipment attached to a wall. During the project, the site was given the name 'Alekša Square', and a large sign of it was attached to the abandoned flag poles that stand in the middle of the square. The bright red color was suggested by the local residents, and it represents both the vibrant and bold community as well as the meaning of the name of the neighborhood ('sarkans' means 'red' in Latvian). Alekša Square was set up collectively, with local residents taking part in site preparation and painting work.

Credits and Data

Project title: Alekša Square
Location: Sarkandaugava neighbourhood, Riga, Latvia
Year: 2013–2014
Urban design: Fine Young Urbanists (Evelīna Ozola)
Landscape design: Ilze Rukšāne,
Vertical gym: Design Catering (Rihards Funts, Kristaps Grundšteins)
Initiators: Contemporary Architecture Information Center (Gunita Kuļikovska), Ideju Māja (Mārcis Rubenis), Sarkandaugavas attīstības biedrība (Alija Turlaja)
Partners: Riga 2014 European Capital of Culture
Construction: Balta Trend, Gumi Mix Group, Design Catering, Direct

亚历山大广场是里加的第一个公共广场，是与萨尔坎多加瓦“问题街区”的本地居民共同设计的。遵循特殊的方法，当地居民先提出想法并在几次会议上讨论关于新公共空间的必要性，随后设计师再将他们的建议和关于必要性的讨论融入广场的设计中。网格状的平面规划构成了区域的基本结构，同时能够根据不同元素带来无限的变化。目前，广场上设有一张乒乓球桌、一个棋盘游戏区、一个休息区以及一个垂直健身区（固定在墙上的定制健身器材）。在项目中，这个地方被命名为“亚历山大广场”，广场中央的废弃旗杆上还挂起了彩旗以作为标志。彩旗使用了亮红色，这来自于当地居民的建议。它象征着充满活力与勇气的社区，也暗指了社区名称的含义（拉脱维亚语中，“sarkans”是“红色”的意思）。亚历山大广场是大家共同建造的成果，场地的整修和涂饰也有当地居民的参与。

pp. 62–63: View of the square from the north. Photo by Ansis Starks, courtesy of the architect.

第 62-63 页：从北边看广场。

Zaiga Gaile Office

Renovation of the Bergs Bazaar

Block among Elizabetes, Marijas and Dzirnavu Str, Riga 1993-

扎格·盖雷建筑师事务所

贝格斯集市改造

里加，伊丽莎白大街、玛丽哈斯大街和兹尔纳乌大街之间的街区 1993-

05

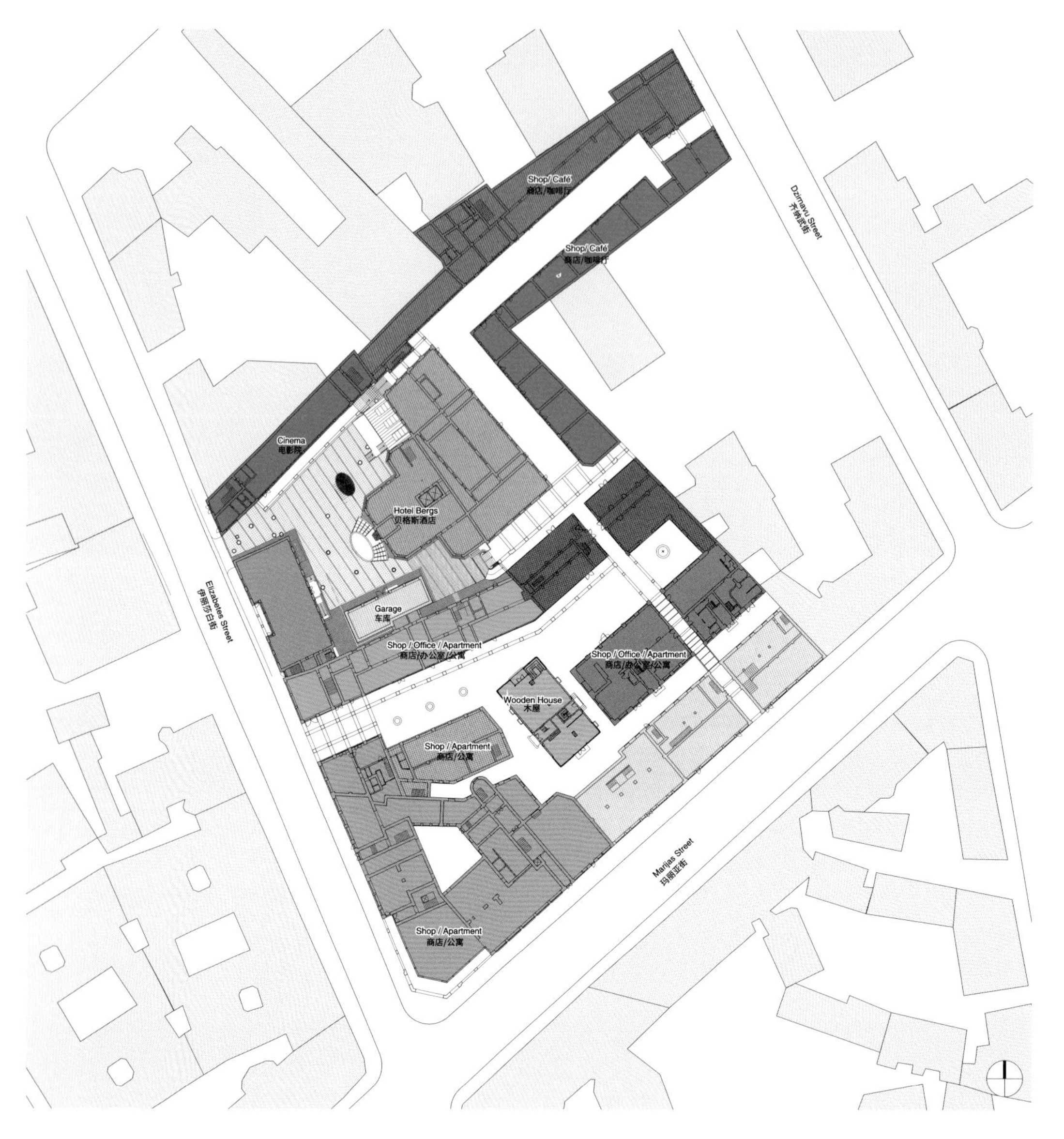

Site plan (scale: 1/1,500)／总平面图（比例：1/1,500）

Credits and Data

Project title: Renovation of the Bergs Bazaar
Client: Justs Karlsons
Location: Block between Elizabetes, Marijas and Dzirnavu streets, Riga, Latvia
Project dates: Started in 1993, still in progress
Construction: Started in 1993, still in progress
Architect: Zaiga Gaile Office
Project team: Zaiga Gaile, Iveta Cibule, Liene Griezīte, Ingmārs Atavs, Ģirts Kalinkevičs, Andra Šmite, Indra Ķempe, Zane Dzintara
Total area of the site: 11,823 m²

pp. 64-65: View of the arcade and plaza. Bergs Bazaar was built as the first shopping mall in Riga at the end of 19th century. After an interval in which it was closed, it now has a cinema, a hotel, restaurants, cafes, offices and apartments. This page, above: Advertisement of Bergs Bazaar from 1898. Courtesy of the architect. This page, below: Aerial view of plaza between cafés and hotel. Photo by Ainars Meiers, courtesy of the architect.

第 64-65 页：拱廊与广场。贝格斯集市是里加市第一家购物中心，建造于 19 世纪末。它曾经历过一次停业，现在拥有一家电影院、一家酒店和众多餐厅、咖啡馆、办公室以及公寓。本页，上：1898 年贝格斯集市的广告；本页，下：俯瞰咖啡馆和酒店之间的广场。

[comfort zone]

Renovation of the Bergs Bazaar is a long-term development project of a whole block in the city center. It started in 1993 as soon as the historical property – the real estate of the Bergs family – was returned to its rightful heirs after 50 years of the Soviet occupation. Archive documents revealed the uniqueness and boldness of the original project – the Bergs Bazaar had been the first retail arcade in Riga. The block was built at the end of the 19th century on a former sand road network and cabbage fields.

The aim of the project was to renovate the unique site according to the original vision of its founder Kristaps Bergs (1843–1907) and the original project of its architect Konstantīns Pēkšēns (1859–1928) and develop it as a modern commercial, retail and residential center, a quiet place in the city center with a unique and charming feeling.

As a result of gradual renovation work over the last 20 years the following has been achieved:

– the abandoned courtyards, arcades and galleries have been restored in their original place, making the Bazaar structurally clear and transparent;

– the neglected cluster of buildings has been renovated and boutique shops, small restaurants, art galleries and other public utilities have been introduced within the original retail spaces on the ground floor;

– office spaces and comfortable apartments with roof terraces and balconies have been made on the upper floors of the Bazaar;

– a design hotel (Hotel Bergs) has been built in the middle of the Bazaar by converting two former apartment buildings into the hotel with a glass covered courtyard between them;

– a cinema building has been renovated;

– the oldest buildings in the block – two wooden houses have been preserved and renovated;

– former garage has been converted into a design wine bar;

– a pedestrian zone with trees, shrubs, flowers, design benches and lighting, fountains and art objects has been developed.

The Bazaar has become a popular place for the city dwellers as well as tourists. There are farmers' markets on Saturdays, art festivals, concerts and other events.

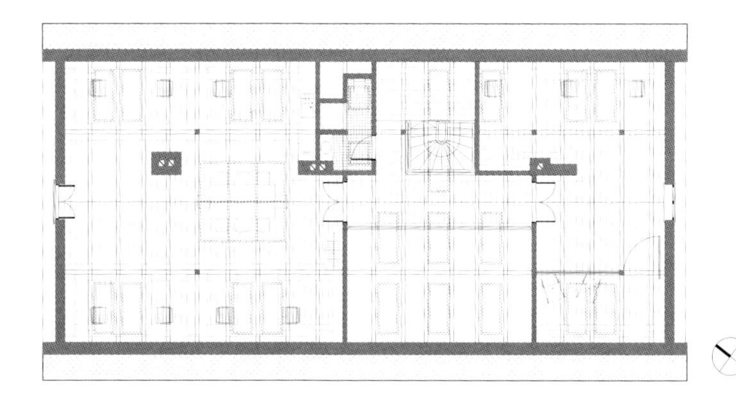

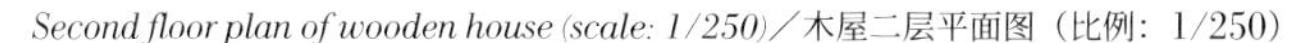

Second floor plan of wooden house (scale: 1/250)／木屋二层平面图（比例：1/250）

Opposite: View toward the west. The wooden house at the center is originally stands on this site and renovated for the office of Zaiga Gaile Office.

对页：向西远望，路中央的木屋为既存建筑，现在被改造为扎格·盖雷事务所的办公室。

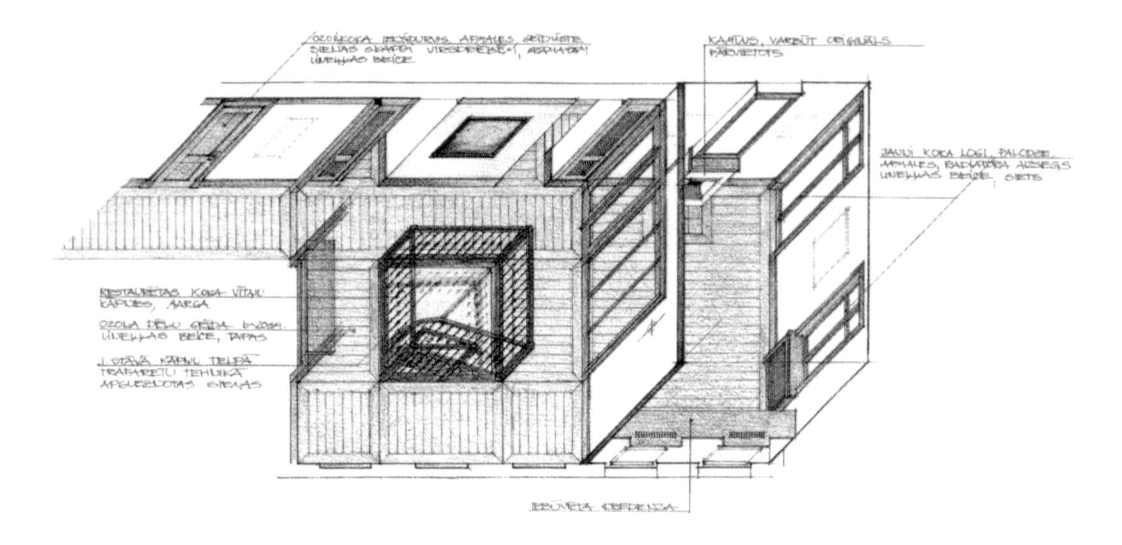

Axonometoric drawing of wooden house／木屋轴测图

贝格斯集市的改造是市中心整个街区的长期开发项目。这里在经历了长达 50 年的苏联时期之后，被归还给贝格斯家族的合法继承人，而改造项目也随之于 1993 年开始。从档案记录中可以了解到，贝格斯集市是里加的第一处零售走廊，这样的规划在当时是独特且大胆的。整个街区建造于 19 世纪末，原先是一片卷心菜地和砂石路网。

这一项目是依据创始人克里斯塔普斯·贝格斯（1843-1907 年）的设想，以及当时的建筑师康士丹斯·帕克辛斯（1859-1928 年）的规划，对这一独特区域进行改造，将其发展为一个现代化的商业、零售和居住中心，和一个在市中心闹中取静、独具魅力的地方。

经过过去 20 年的逐步改造，项目已经实现了：

- 在原址上修复已被废弃的庭院、拱廊及画廊，使得集市的结构更加清晰、明确；

- 翻新被忽视的建筑群，并吸引时装店、小餐厅、艺术画廊及其他公共设施入驻地上层原本的零售空间；

- 在集市的上层建造带有屋顶露台及阳台的办公空间和舒适的公寓；

- 在集市的中央建造一座设计酒店“贝格斯酒店”。将原有的两栋公寓加以改造，并在两栋楼之间设置一座带有玻璃屋顶的庭院；

- 电影院翻修；

- 保留并改建园区内最古老的建筑——两座木屋；

- 将过去的仓库改造为设计酒吧；

- 创造一个有树木、灌木、花草、设计长凳以及照明、喷泉、艺术品的步行区。

改造后的贝格斯集市受到了本地居民和游客的欢迎。这里每周六还会举办农贸市集、艺术节、音乐会及其他活动。

This page: Interior view of the second floor in the Zaiga Gaile Office office. pp. 72–73: Interior view of the second floor office. An unused attic room was transformed into a studio preserving the structure and providing the skylight.

本页：扎格·盖雷建筑师事务所办公室二楼内景。第 72-73 页：办公室二楼内景。一间闲置的阁楼被改造为工作室，既保留了原有结构，也新设置了天窗。

Eden

Sudraba Arhitektūra

Staircase Renovation Art Deco

Elizabetes Str. 21a, Riga 2013

苏德拉巴建筑师事务所

装饰艺术楼梯间改造

里加，伊拉莎白大街21a号 2013

Site plan (scale: 1/8,000) ／总平面图（比例：1/8,000）

9

Staircase interiors in Riga apartment houses built at the turn of the 19th and 20th centuries were mostly splendid, rich in decoration, as a continuation of the facade. This apartment building with commercial spaces, located at the intersection of four streets, was designed by architect Martins Nuksa and also had a luxurious entrance hall. In the facade of this neoclassical building columns with Corinthian capitals support a frieze with motives of antique art while the horizontal lines of the windows and decorative elements outline the approaching Art Deco and functionalism. The essence of staircase is created by spiral stairs and lantern-glass cupola. In the 1930s, the Latvian Rē rihs Society was located in this building.

Unfortunately after a renovation carried out in the 1980s most of the original staircase and entrance hall finishing details were lost. Only the spiral staircase was still in place.

During careful investigations, a fragment of a 35-cm high line with ornament in stencil technique was found under the door casing. This ornamental line initially had been decorating all walls in the staircase. The ornament consists of palmette motives (popular in Ancient Greek art) and neoclassical flower ornamentation. This combination corresponds to the aesthetics of the building's facade. From this discovery derived an idea to turn this ornament into a three-dimensional element and integrate it into the interior finishing of the staircase. This three-dimensional ornament pressed into plaster as well as stained glass skylight became main elements of renovated interior. Concrete floor covering plates and skirting boards were produced in Latvia especially for this project.

Credits and Data
Project title: Staircase Renovation Art Deco
Location: Elizabetes Str., Riga
Realization: 2011–2013
Architects: Sudraba Arhitektura / Reinis Liepins, Ilze Liepina
Stained glass skylight: Artis Nimanis (An&Angel)

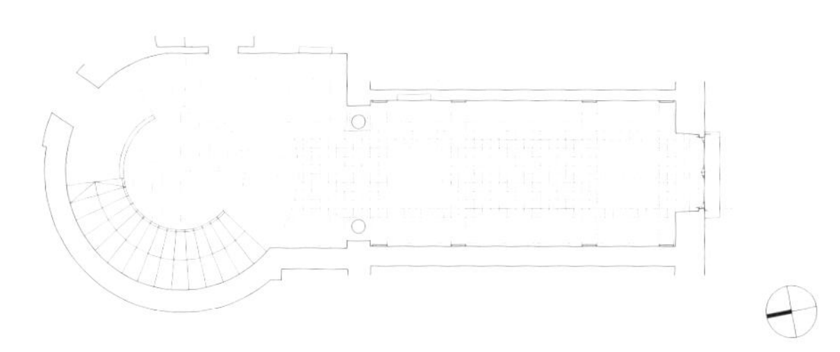

Ground floor plan (scale: 1/250) ／一层平面图（比例：1/250）

p. 74: Evening view from the west. Photo by Didzis Grodzs courtesy of the architect. p. 75: View of the renovated staircase. All photo on pp. 74–79 expect as noted by Ansis Starks courtesy of the architect. Opposite: Looking up to the skylight. Stensile pattern is applied below the skylight. p. 79, above: Interior of the lobby hall. p. 79, below left: Palmette and neoclassicist flower pattern found in a fragmentary surviving ornament. p. 79, below right: Rubber mold to stamp new pattern based on the surviving pattern. Two photos by Reinis Liepins.

第 74 页：从西侧看到的夜景。第 75 页：改造后的楼梯间。对页：仰视天窗。天窗下方有压印的装饰模块。第 79 页，上：大厅内景；第 79 页，左下：在遗址中发现的棕榈叶与新古典主义花卉装饰残片；第 79 页，右下：以残片花纹为原型设计的新图案，并利用橡胶板进行压印。

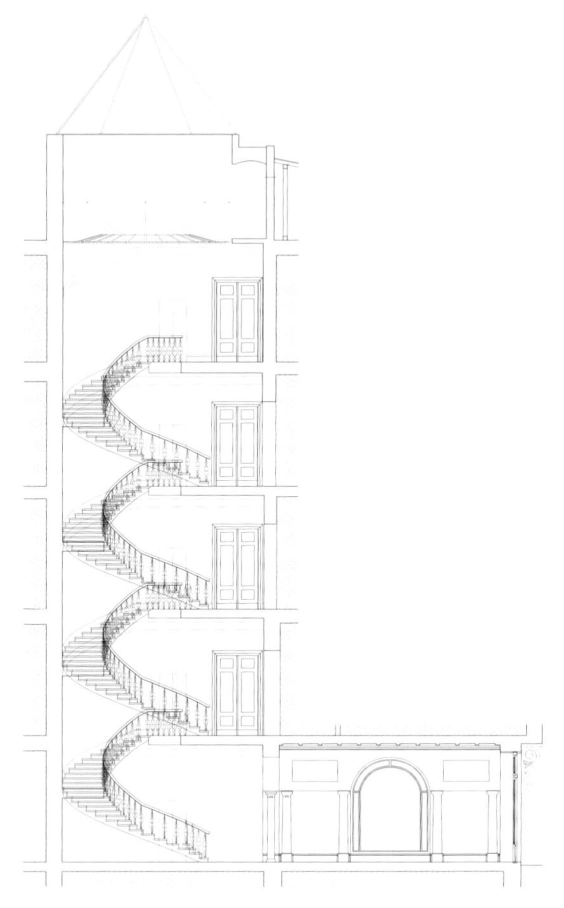

Section (scale: 1/250)／剖面图（比例：1/250）

里加 19 世纪末到 20 世纪初建造的公寓楼梯间大多风格华丽、装饰丰富，是建筑立面的延续。这座带有商业空间的公寓是由建筑师马蒂斯·努克萨设计的。它位于四条街道的交叉口，并拥有一个华丽的门厅。在这座新古典主义建筑中，科林斯式柱子支撑着描绘有古典艺术主题的饰带。而窗户的水平线和装饰要素则勾勒出装饰艺术及功能主义的特征。这座楼梯间是由螺旋阶梯和玻璃穹顶构成的。20 世纪 30 年代，里加雷利斯协会就位于这座建筑中。

不幸的是，经过 20 世纪 80 年代的一次改建，除螺旋楼梯外，原楼梯间及门厅的装饰细节大多已经消失。

在详细的基地调研中，一块 35 厘米高的装饰碎片被发现于门框之下，上面还带有模块化的装饰。原本整个楼梯间的墙面都被这样的装饰线所装饰，上面的图案包括棕榈叶（在古希腊艺术作品中常见）、新古典主义花卉装饰等。这样的组合与建筑立面的美学相呼应。受这一发现的启发，这种装饰被转化为三维立体元素，整合到楼梯间的内部装饰中。这种被压入墙面灰泥的立体装饰以及彩色玻璃天窗成为改建后楼梯间的主要特色。覆盖有板材和护墙板的混凝土地面是为这一项目专门定制的，其产地就在拉脱维亚。

Sudraba Arhitektūra

House in the Courtyard

Miera Str. 52a, Riga 2008

苏德拉巴建筑师事务所

庭院中的房子

里加，米埃拉大街52a号 2008

07

Building is located on Miera Street, Riga, in a courtyard between two wooden buildings, characteristic of 19th century architecture in Riga. Building volume and layout outline with rounded corners correspond to the building from 1950's, which was impossible to preserve due to technical condition of constructions. The first floor of building is created as a relief concrete base on top of which two-storey wooden volume is located. This solution was determined by context of surrounding wooden buildings as well as functional differences of volumes (first floor and basement floors as office premises, second and third floors as apartments). To provide necessary amount of daylight in all apartment areas, continuous staircase is located at building's firewall. Building is completed by fourth floor roof volume with terrace.

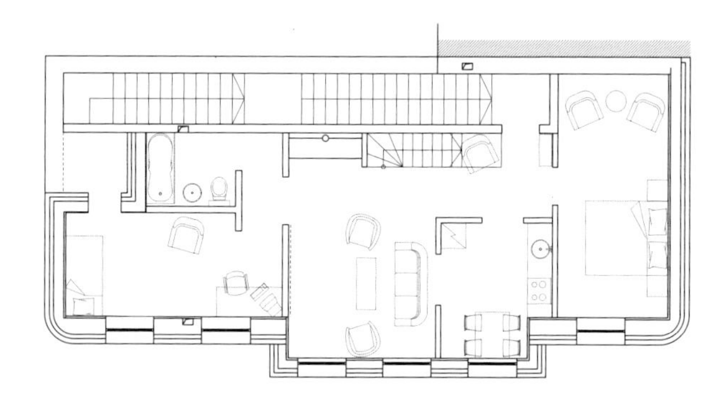

Third floor plan /三层平面图

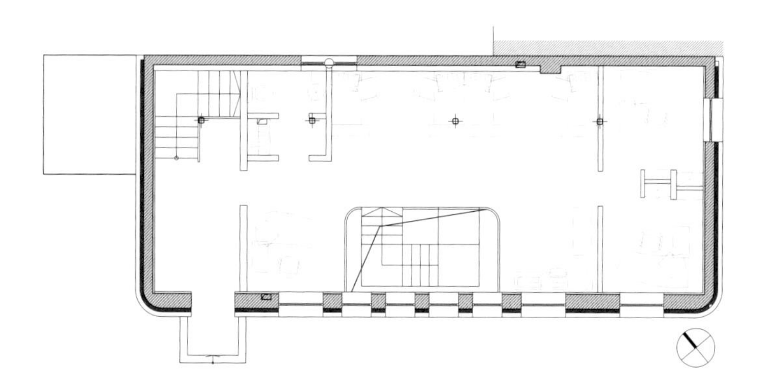

Ground floor plan (scale: 1/200)/一层平面图（比例：1/200）

Credits and Data

Project title: House in the Courtyard
Location: Miera Str. 52a, Riga, Latvia
Completed: 2008
Architect: Sudraba Arhitektura / Reinis Liepins
Construction engineer: IG Kurbads

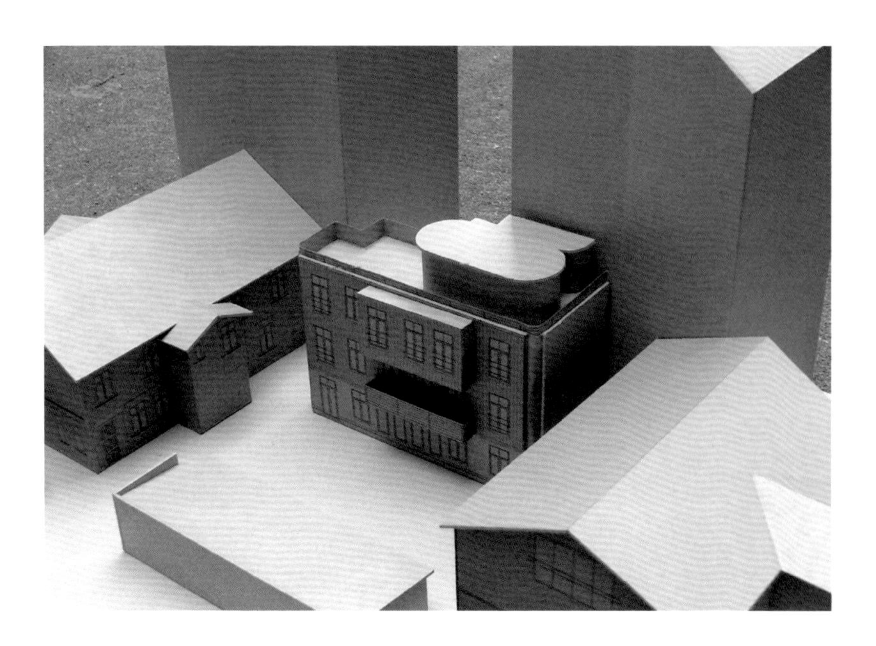

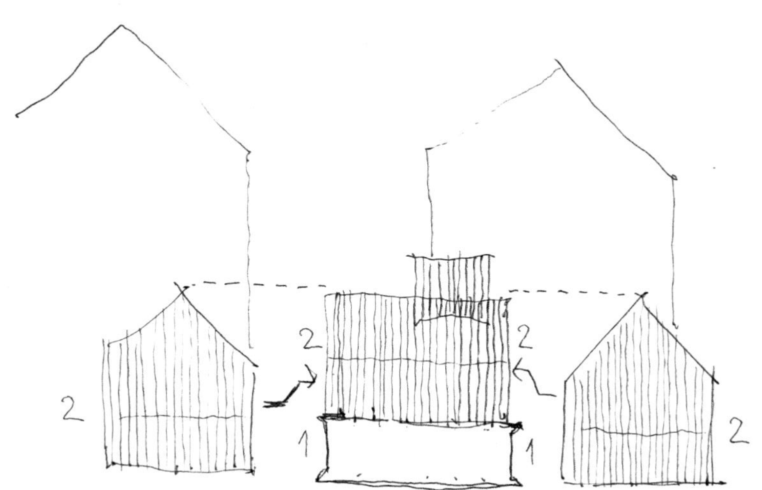

Sketch／手绘草图

pp. 80–81: General view of the north side. The house stands on the typical courtyard of wooden buildings in Riga. This page: Site model. Image courtesy of the architect.

第 80-81 页：北侧全景。这座房子坐落在里加典型的木造建筑庭院中。本页：场地模型。

这座建筑位于里加的米拉街，坐落在两座具有里加 19 世纪建筑特色的木造建筑物之间。这座圆角建筑的体量及平面轮廓与 20 世纪 50 年代的建筑相呼应，不过由于建造技术的限制，当时的建筑未能保存下来。建筑一层是利用浮雕混凝土建造的，其上还有两层木质建筑。这一方案是由周围的木质建筑环境及建筑内的功能差异（建筑一层及地下室是办公场所，二层和三层是公寓）所决定的。为了让公寓中的所有区域都能获得必要的阳光，连续的楼梯被设置在建筑的防火墙处。建筑顶层，也就是四层是屋顶和露台。

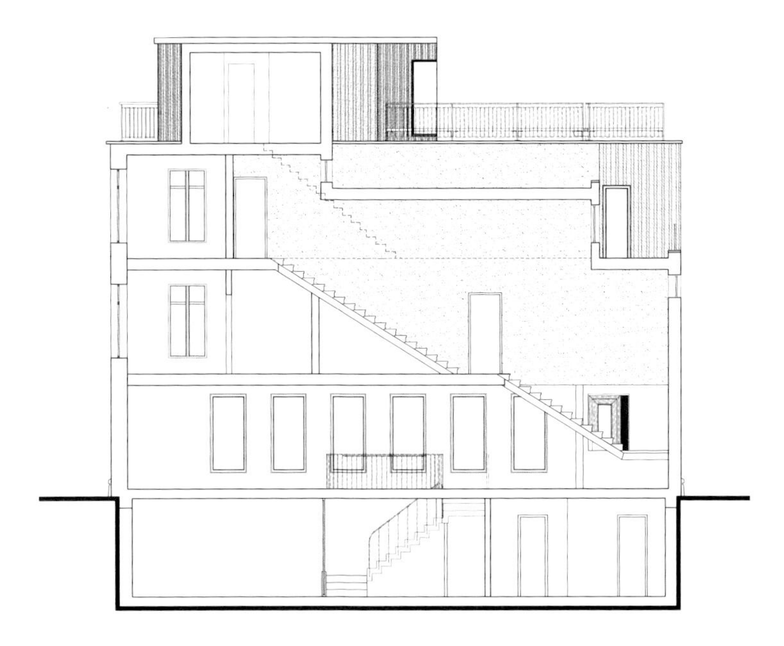

Section (scale: 1/200)／剖面图（比例：1/200）

Opposite, above: View from the terrace. Opposite, below: Interior of the office on the third floor.

对页，上：在露台上眺望；对页，下：三楼办公室的内景。

Arhis

Office and Retail Premises Building in Riga, Baznīcas street 20/22

Baznīcas Str. 20/22, Riga 2003

阿尔希斯建筑师事务所

里加巴兹尼卡斯大街20/22号办公兼商业大楼

里加，巴兹尼卡斯大街20/22号 2003

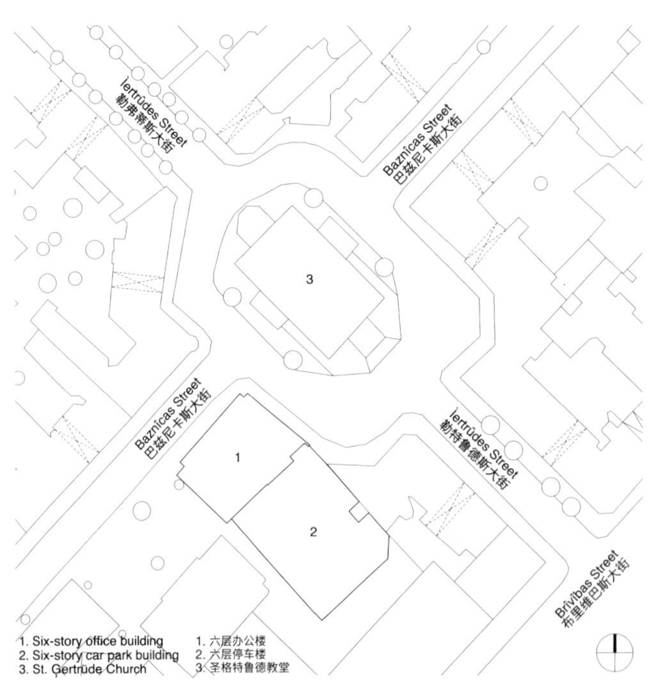

Site plan (scale: 1/2,000)／总平面图（比例：1/2,000）

Credits and Data

Project title: Office and Retail Premises Building in Riga, Baznīcas street 20/22
Project: 1999–2000
Construction: 2003
Architects: A. Kronbergs, E. Beernaerts, B. Bula, V. Šlars
Architect technicians: J. Vizulis, Ē. Geinbergs, R. Saulītis, A. Staris

iznomā veikalu un biroju telpas
DOMUSS
iznomā veikalu un biroju telpas
DOMUSS

The spatial development helps to continue the perimeter structure and emphasizes the location of the building on two landed estates. The first and second floors of the building are occupied by retail premises and the third to sixth floors by office spaces with access to garage floors.

The building is made of monolithic reinforced concrete. Facades are formed as suspended heat non-insulated glass planes. The actual walls are heat insulated reinforced concrete planes with multi-level wooden trim decoration.

空间的展开有助于保持周边构造的连续性，还能够强调建筑在两块地基上的位置。大楼的一层和二层是商业零售场所，三层到六层为办公空间并通向停车楼层。

整栋大楼是以整体式钢筋混凝土建成的。外墙用悬挂式非隔热玻璃板建造，而实体墙则使用了隔热钢筋混凝土板，其上有精致的多层木质装饰。

p. 87: General view from the north. Behind the double skin of glass, the building has a wooden structure. St. Gertrude Church is seen on the left, and apartment built in 1918 is seen on the right. Photo by I. Stūrmanis.

第 87 页：自北面看到的全景。建筑的双层玻璃之后是木制结构。圣格特鲁德教堂位于建筑的左侧，建于 1918 年的公寓位于建筑的右侧。

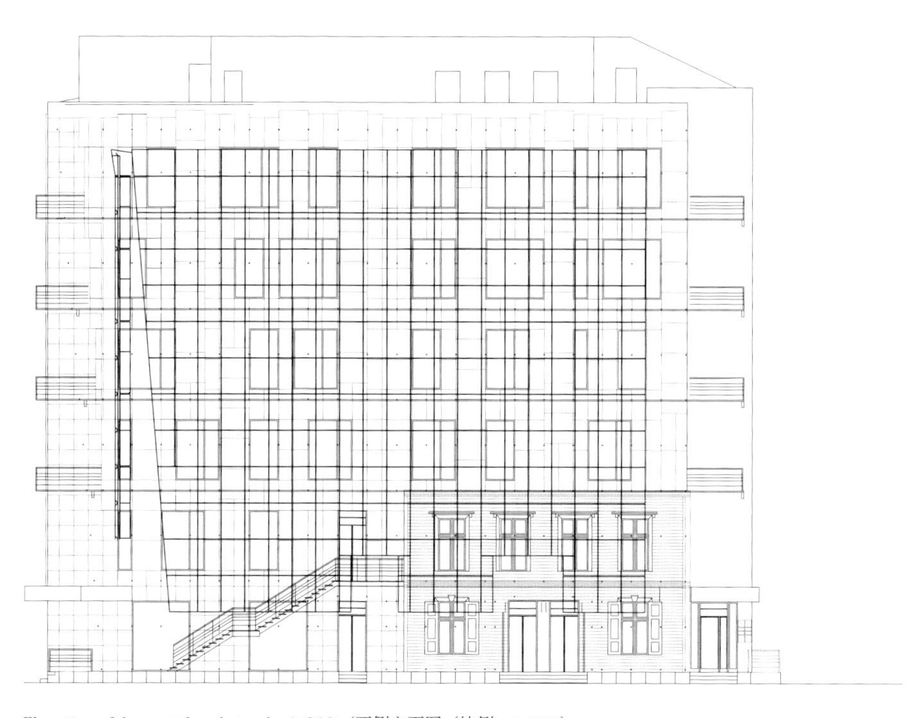

Elevation of the west facade (scale: 1/300)／西侧立面图（比例：1/300）

Zaiga Gaile Office

New Riga Theatre Reconstruction Project

Lāčplēša Str. 25, Riga 2014–

扎格·盖雷建筑师事务所

新里加剧院改造

里加，乐普萨大街25号 2014-

09

Zaiga Gaile Office and partner won the competition for reconstruction of the New Riga Theatre in 2014. The project includes reconstruction of three historical buildings and design of a new volume.

The aim of the reconstruction is to preserve as much of the historical essence and character of the original buildings as possible and delicately add new architecture and facilities of the 21th century.

The project proposes to make several gentle alterations to the main historical facade both for design and technical reasons which all together will make the theatre building stand out as a prominent cultural institution within the urban structure. The original theatre building has never had a proper front area – it is positioned right next to a busy and noisy street. The reconstruction project proposes to relocate the main entrance from the street to the courtyard. The courtyard will be accessed through a passage tunnel on the left side of the building.

The new entrance is located at the far end of the courtyard. Behind the glass wall and revolving door there is a lobby with a ticket office. The courtyard and the lobby are united with the same cobblestone flooring. In the same way the brick walls of the old houses are also exposed in the interior of the lobby. In the center of the lobby by the old brick wall are placed monumental stairs to the first floor. The glass foyer on the first floor with the public balcony above the entrance is the main space for the audience to spend intermissions. In the foyer next to the staircase is a counter with refreshments, and tables are arranged by the glass wall with a view to the courtyard. Under the stairs on the ground floor in the center of the lobby is a fireplace which has several symbolical connotations and historical references. Both the historical hall and the new halls can be easily accessed from the central lobby. All halls have their independent lobbies with wardrobes and facilities.

As a result of the project the theatre will get a gently reconstructed historical hall with 500 seats, and two new halls with 250 and 100 seats, all equipped with the newest technical facilities.

Credits and Data

Project title: New Riga Theatre Reconstruction Project
Client: State Stock Company Valsts Nekustamie īpašumi
Location: Lāčplēsa street 25, Riga, Latvia
Project dates:2014 –present
Architect: Zaiga Gaile Office and Partner
Project supervisor: Zaiga Gaile
Project team: Zaiga Gaile, Ingmārs Atavs, Dāvis Gasuls, Agnese Sirmā, Filips Pitens, Uldis Eglītis, Maija Putniņa- Gaile, Sintija Norberte, Kristīne Riba
Chief engineer: Māris Alsiņš
Construction: Kurbads / Normunds Tirāns
Theatre technology: Theateradvies bv / Louis Janssen
Acoustics: Margriet Lautenbach
Investigation: Balts un Melns / Māris Alsiņš

JAUNAIS RĪGAS TEĀTRIS
GA MĀJA
GERRY WEBER
EBER

扎格·盖雷建筑师事务所在 2014 年新里加剧院改造项目的竞赛中获胜。整个项目包括三座历史建筑的改造以及一座新建筑的设计。

改造目的是尽可能地保留原有建筑的历史本质和特征，同时巧妙地增添 21 世纪新的建筑特点和设施。

出于设计和技术的需要，项目对主要历史建筑的外立面进行了一些细微的调整。这使得整个剧院建筑成为城市构成中一个重要的文化机构。由于剧院坐落在一条繁忙且嘈杂的街道旁，它过去并没有一个合适的前厅区域。我们提出将主入口从道路一侧移到中庭，而通过建筑左侧的通道可以进入中庭。

新入口位于中庭尽头。在玻璃墙和旋转门后面是一个设有售票处的门厅。中庭和大厅的地面使用了同样的鹅卵石，以保持其整体性。同样，大厅内部裸露出旧房子的砖墙。其中心靠近旧砖墙的地方设置有通向二楼的坚固楼梯。二楼，就在入口的正上方处，设有玻璃门斗及公共阳台，这是观众幕间休息时的主要休闲空间。楼梯旁的门斗中有一个提供小食的柜台以及许多桌子。这些桌子靠着玻璃墙布置，坐在桌边可以看到中庭。在一层门厅的中央、楼梯的下方有一个壁炉，它有一定的象征意义和历史价值。从中央门厅可以很容易地进入新旧乐厅。所有的乐厅都有独立的门厅、衣帽间及其他设施。

作为项目的最终结果，这一剧院的旧乐厅将被逐渐改造，并拥有 500 个座位。而两个分别设有 250 个和 100 个座位的新乐厅也将被建立。它们都将配备最新的技术设施。

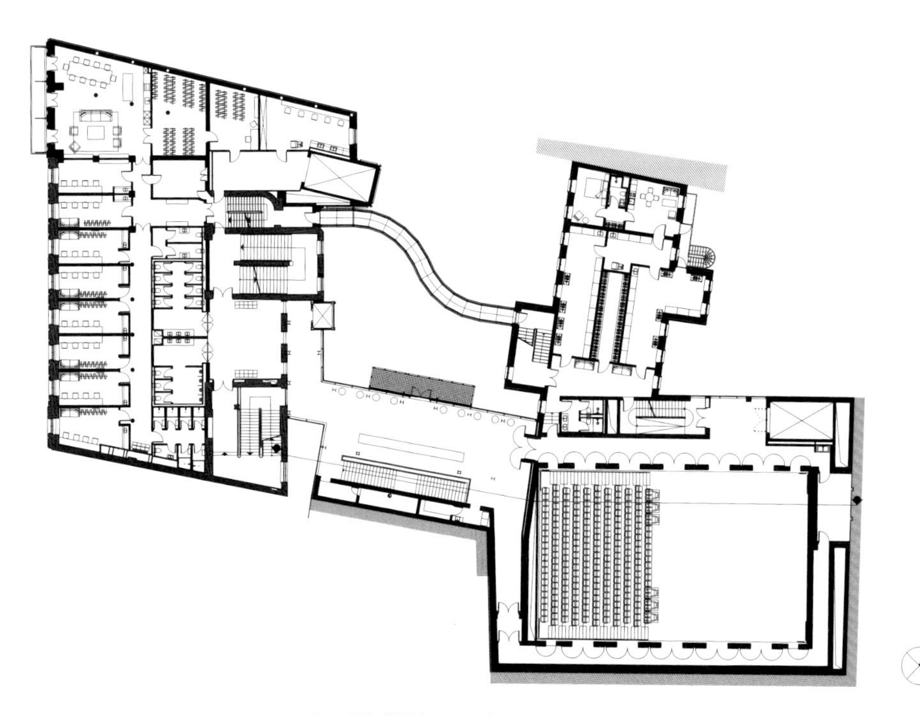

Second floor plan (scale: 1/500)／二层平面图（比例：1/500）

p. 91: Image of the courtyard. It is surrounded with historical yellow brick houses. The reconstruction adds glass panels and industrial metal constructions. The space is designed for people to meet and enjoy the moment. Opposite: Exterior view from the Lāčplēša street. The building is located in the city center among residential buildings. It was originally built for the Society of Craftsmen and has housed the theatre for almost one hundred years. The entrance on the left leads to the courtyard.

第 91 页：庭院景观。整个庭院被历史悠久的黄砖房所围绕。改造增加了玻璃板及工业金属结构。整个空间是为人们的聚会和休闲而设计的。对页：自乐普萨大街看到的外观。这座剧院建造在市中心的住宅楼之间。它最初是为工匠协会而建的，被作为剧院使用已有近百年。左边的入口通向庭院。

This page: View of the roofscape of boulevards and streets in the center of Riga, which was mostly built up at the beginning of the 20th century. Architectural styles vary from Neoclassical style to Art Nouveau (Jugendstil), Art Deco and Functionalism. At the back left, Riga Radio and TV Tower (1979–1989) and at the right the Latvian Academy of Sciences (1951–1961), both being urban heritage from the Soviet regime period, are seen. Opposite: Close up view of the travertine facade.

本页：里加市中心街道上的屋顶全景。他们主要建于 20 世纪初期。建筑风格从新古典主义到新艺术、装饰艺术和功能主义，丰富多变。在照片的左后方可以看到里加广播电视塔（1917-1989），右边可以看到拉脱维亚科学院（1951-1961）。它们都是苏联执政时留下来的城市遗产。
对页：石灰立面特写。

Apartments and Retail Premises
Aleksandrs Klinklāvs
Brīvības Str. 40
1934

公寓和零售场所
亚历山大·克林克洛夫
布里维巴斯大街40号
1934

10

Atis Ķeniņš School
Konstantīns Pēkšēns, Eižens Laube
Terbatas Str. 15/17
1905

阿提斯·肯尼思学校
康士丹斯·珀格、艾森·劳贝
塔伯特斯大街15/17号
1905

pp. 96-97: Street view of Tērbatas Str. from the northeast. The building is one of the first examples of National Romanticism in the Art Nouveau architecture of Riga.

第96-97页：从东北方向看塔伯特斯大街的街景。这座建筑是里加新艺术运动民族浪漫主义建筑的早期实例之一。

公寓

艾森·劳贝

阿尔伯特大街11号

1908

This page: General view from the east. Opposite: View of the lavishly decorated entrance of the Art Nouveau apartments building, now housing the Art Nouveau Museum at Alberta Str. 12. Central spiral staircase is seen in the back.

本页：从东侧看到的全景。对页：新艺术风格公寓装饰华丽的入口，现在是新艺术博物馆，位于阿尔伯特大街 12 号。建筑内还可以看到中央螺旋楼梯。

Apartments (Riga Art Nouveau Museum)
Konstantīns Pēkšēns, Eižens Laube
Alberta Str. 12
1903

公寓（里加新艺术博物馆）
康士丹斯·珀格、艾森·劳贝
阿尔伯特大街12号
1903

13

Apartments and Retail Premises
Mihaels Eizenšteins
Elizabetes Str. 10a
1903

公寓兼商业设施
米哈伊尔·爱森斯坦
伊丽莎白大街10号a座
1903

14

Dailes Theatre
Marta Staņa, Imants Jekabsons, Haralds Kanders
Brīvības Str. 75
1959–1976

戴尔斯剧院

马塔·斯塔纳，伊曼兹·雅各布森，哈拉茨·坎德尔斯

布里维巴斯大街75号

1959-1976

15

Opposite: View of the facade from the northwest. This page: General view from the Bruņinieku street.pp. 102–103: View towards the interior of the theater to the Bruņinieku street.

对页：从西北方看建筑立面。本页：从布鲁尼克大街看到的全景。第 102-103 页：从剧院内看布鲁尼克大街。

Old Town of Riga

里加老城

Being part of the Hanseatic League, Riga was one of the flourishing urban centers of Northeast Europe in the 13th–15th centuries. Although most of the earliest buildings were destroyed by fire or war, the current urban street pattern still induces the atmosphere of the irregular winding streets of Medieval Riga. Over its long history, the Old Town of Riga has undergone multiple structural changes and reconstructions, currently posing the questions for architects to find a seamless balance between dense historical substances and contemporary interventions.

作为汉萨同盟的一部分，里加是13–15世纪欧洲东北部繁荣的中心城市之一。虽然这里的早期建筑大多因火灾或战争遭到毁坏，但如今的街道格局仍然留存着中世纪里加不规则蜿蜒街道的感觉。在漫长的历史发展中，里加老城经历了多次结构性变化和重建，这也给建筑师们提出了一个难题：如何在密集的历史遗产和现代介入之间实现无缝衔接。

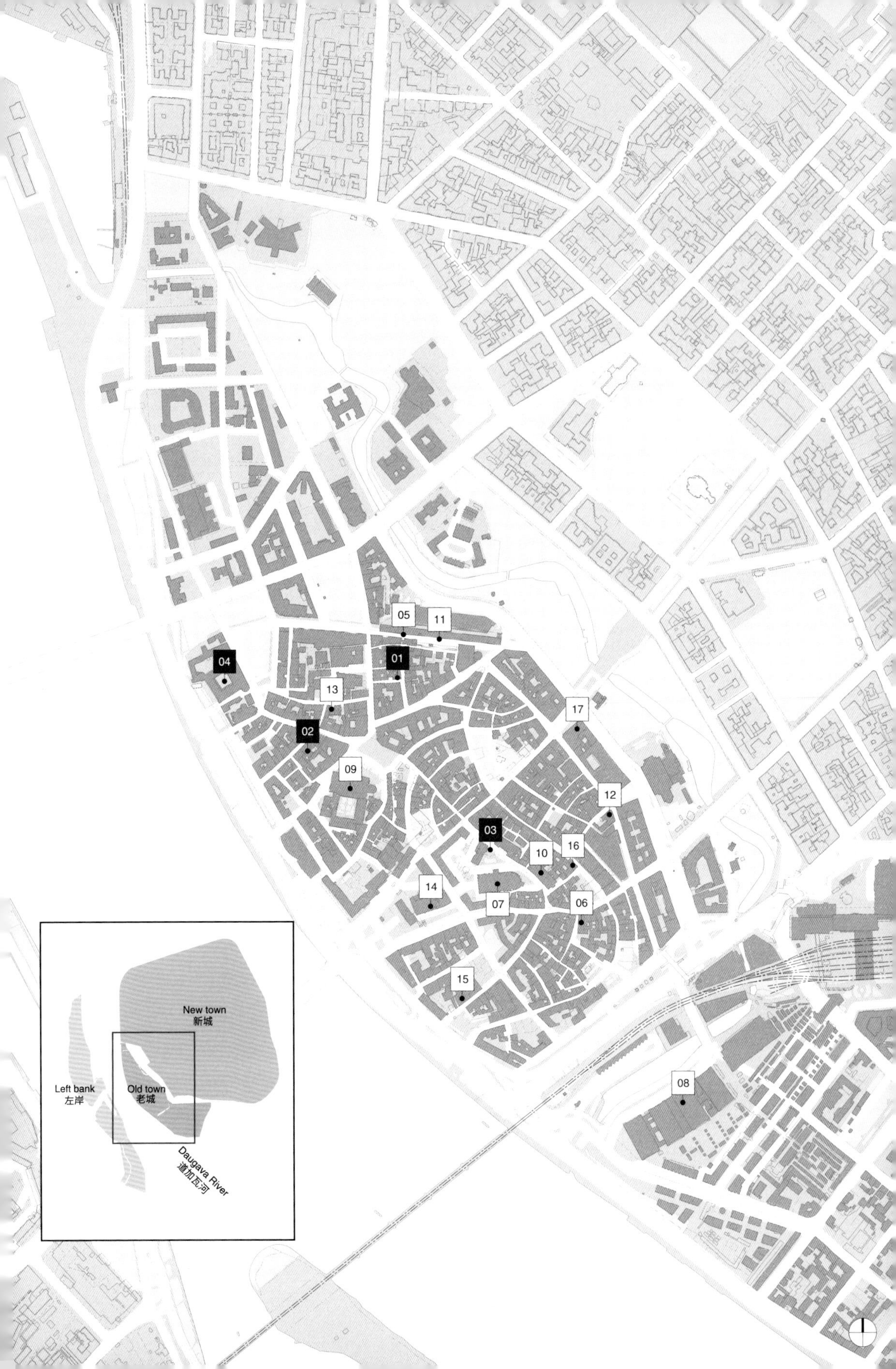

05
11
04
01
13
17
02
09
12
03
10
16
14
07
06
15
08
New town
新城
Left bank
左岸
Old town
老城
Daugava River
道加瓦河

Renovation of 17th Century Warehouse building in Riga
Sudraba Arhitektūra, Arhitektes Lienes Griezītes studija
Aldaru Str. 5, Riga/ 2011 (pp. 108–117)

"Dome Hotel & SPA", Conversion of 17th Century Apartment Building
Sudraba Arhitektūra
Miesnieku Str. 4, Riga
2009 (pp. 118–125)

Residential Building on Skārņu Str. 11, Riga
Jaunromāns un Ābele
Skārņu Str. 11, Riga
2014 (pp. 126–131)

Renovation Project of the Castellum of Riga Castle
Sudraba Arhitektūra, MARK arhitekti
Pils Square 3, Riga
2014–2020 (pp. 132–137)

Sweedish Gates and Fortification Wall
Torņu and Aldaru Str. corner
13 C. (Wall), 1698 (Gates).
(p. 140-141)

Apartments and Retail Premises
Alfrēds Ašenkampfs, Maksis Šervinskis
Audēju Str. 7
1899 (p. 142-144)

St. Peter's Church
Skārņu. Str.19, Kaļķu Str. 2 and Grē cinieku Str.11
13–19th Century, 1968–1977 reconstruction(p. 145-147)

Riga Central Market
Pāvils Dreijmanis, Pāvils Pavlovs, Vasilijs Isajevs, Georgs Tolstojs
Nēģu Str. 7
1924–1930 (pp. 148–149)

Riga Cathedral
Herdera Sqr. 6
13–19 C

St. John's Church
Skārņu Str. 24
13–17 C

Reconstructed Jēkaba Kazarmas, reconstructed City Fortification Wall with Ramer Tower
Torņa Str.
13 C, 16–17 C, 1989

Apartments and Retail Premises
Heinrihs Šēls, Frīdrihs Šēfels
Teātra Str. 9
1903

Apartment Buildings ensemble "Three Brothers"
Mazā Pils Str. 17, 19, 21
15 C, 17 C, 20 C

House of the Blackheads reconstruction
Uģis Bratuškins, Ināra Dzene, Ina Kuļ ikovska, Ēriks Zīle, Edgars Pučiņš, Vija Caune, Ināra Batraga / Rātslaukums 7 / 14 C, 16–17 C, 19

House of Dannenstern
Rupert Bindenschu
Mārstaļu Str. 21
1696

Apartments and Retail Premises
Pauls Mandelštams
Kalēju Str. 23
1903

"Hotel de Rome"
Edvīns Vecumnieks, V. Zilgalvis, Dainis Bērziņš, Jānis Kārkliņš, Ēriks Zīle
Kaļķu Str. 28 / 1987–1992

Sudraba Arhitektūra, Arhitektes Lienes Griezītes studija

Renovation of 17th Century Warehouse Building in Riga

Aldaru Str. 5, Riga 2011

苏德拉巴建筑师事务所，莱恩斯·格里兹建筑师事务所

里加17世纪仓库更新

里加，阿尔达鲁街道 5号

2011

01

Brīvdabas
- Kino Vakari -

The warehouse building, a 17th- century architectural monument, appears to be one of the rare Riga downtown buildings that has retained the inspiration of its initial origin. In the course of time the building has obtained one side inclination, thus reinforcing its individuality. As a result of reconstruction, the six-level building (777 m²) was transformed into a universal space open for the city. The lower floors are fit for art galleries and stores, the next two for special Riga downtown offices with low ceilings, and the attic floors for two-level apartments with small tile roof terraces. The building can be managed as a functionally unified aggregate.

A transparent elevator and robust, dark stairs were inserted in the center of the building. The initial stone wall and wooden constructions are kept intact with the characteristic historical overlay. The architecture of the new interior respects the historical structure of the building, while at the same time being in contrast to it.

Despite the seemingly introvert air about the building and invisibility of renovation form the outside, we strove to find an organic link with the city. Ground floor has seen most transformations: original floor height was preserved at the room's deepest end, forming a paved street – level gallery along outside wall.

Credits and Data

Project title: Renovation of 17th Century Warehouse Building in Riga
Location: Aldaru street 5, Riga, Latvia
Completed: 2011
Architects: Sudraba Arhitektūra, Arhitektes Lienes Griezites Studija
Project team: Liene Griezite, Liga Apine, Reinis Liepins

pp. 108–109: Exterior view from the east. It is located on a street of warehouses facing the Swedish Gate. While preserving the facade and structure, the reconstruction transformed the warehouse into offices and apartments. Opposite: View of the elevator and stairs. Modern circulation was added to match contemporary lifestyles.

第 108-109 页：从东侧看外观。这座仓库位于一条面向瑞典门的仓库街上。在保留原立面和结构的基础上，改造将仓库转变为办公空间和公寓。对页：电梯和楼梯。为适应当代生活方式，现代化的动线被加入进来。

这座仓库是里加17世纪建筑的纪念碑。它位于里加市中心，是这里少有的、保留有最初设计的建筑之一。随着时间的流逝，这座建筑逐渐向一侧倾斜，这使得它更具特色。经过改造，这座六层建筑（777平方米）被设计为通用性极高的城市开放空间。低层主要用作画廊和商店，中间两层为层高较低的特别办公室，阁楼是双层公寓，还设置有瓦葺屋顶的阳台。整栋建筑可以被作为功能性集合体来管理。

建筑中央有一部透明电梯和一座坚固的暗色楼梯。原有的石墙和木造结构被保留下来，以呼应这里独特的历史积淀。新的内部空间在尊重历史的同时，也与之形成对比。

这座建筑在风格上看似内敛且从外观上看不到改造的变化，但我们致力于找到它与城市之间的有机联系。一层的改变是最多的：地板高度根据原房间的最低处得以保留下来，从而沿着外墙铺就成一个与街道水平的画廊。

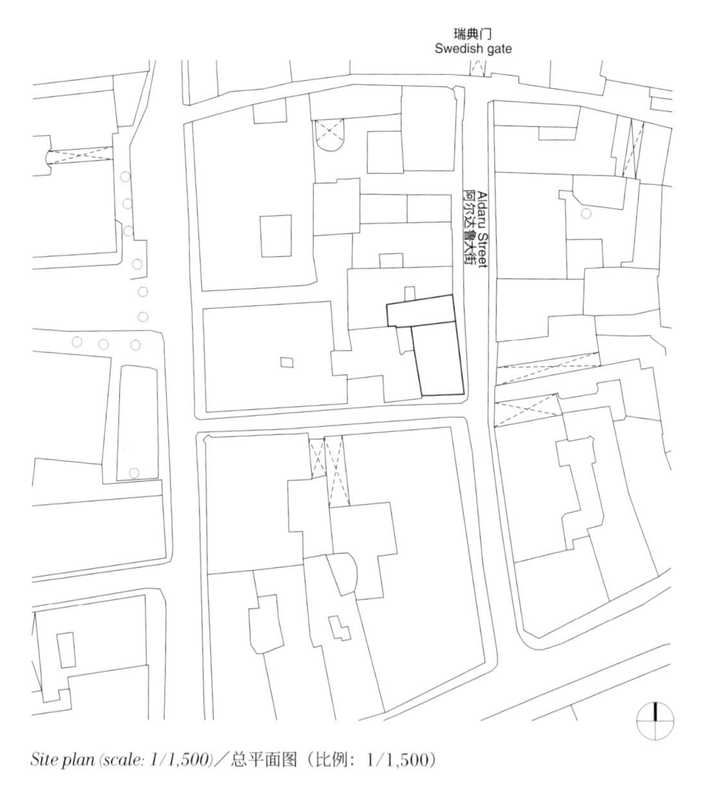

Site plan (scale: 1/1,500)／总平面图（比例：1/1,500）

pp. 112–113: Interior view at the fifth floor. A pulley system to lift goods into the warehouse was preserved.

第 112-113 页：五楼内景。用于将货物吊入仓库的滑轮系统被保留下来。

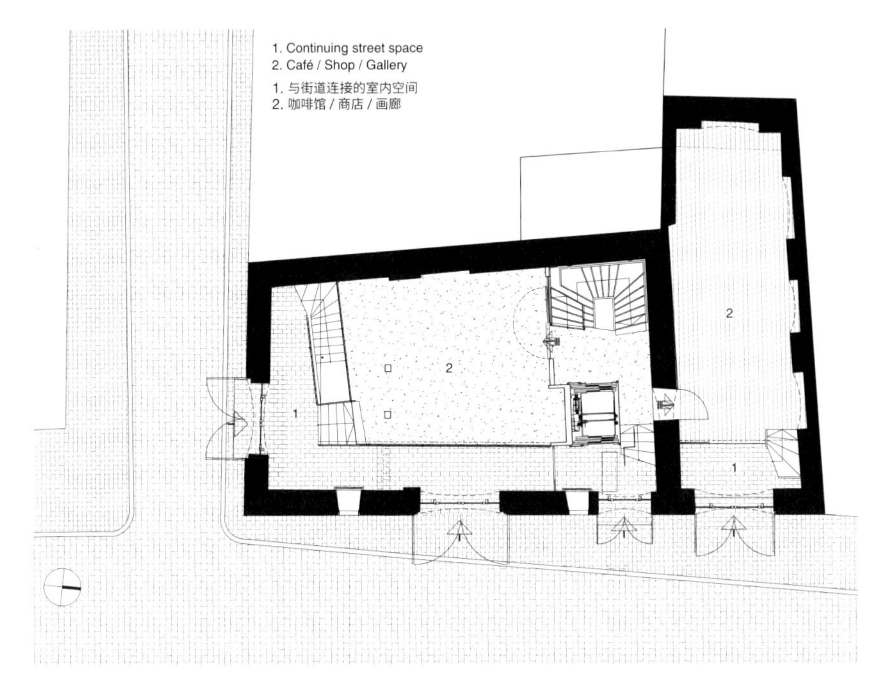

Ground floor plan (scale: 1/250)／一层平面图（比例：1/250）

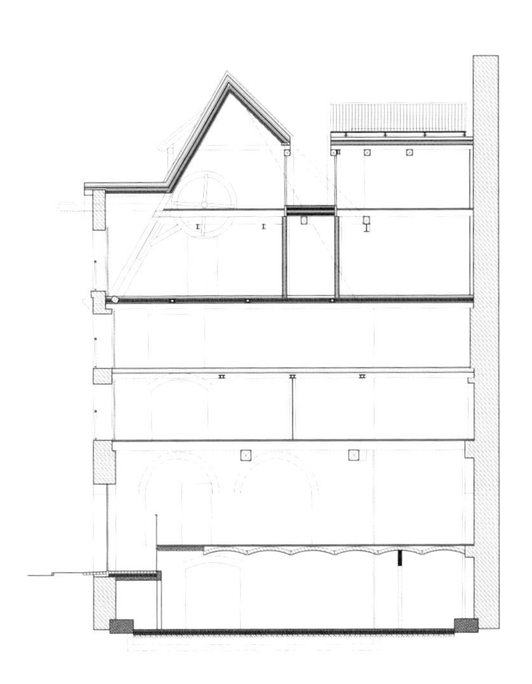

Longitudinal section (scale: 1/250)／纵向剖面图（比例：1/250）

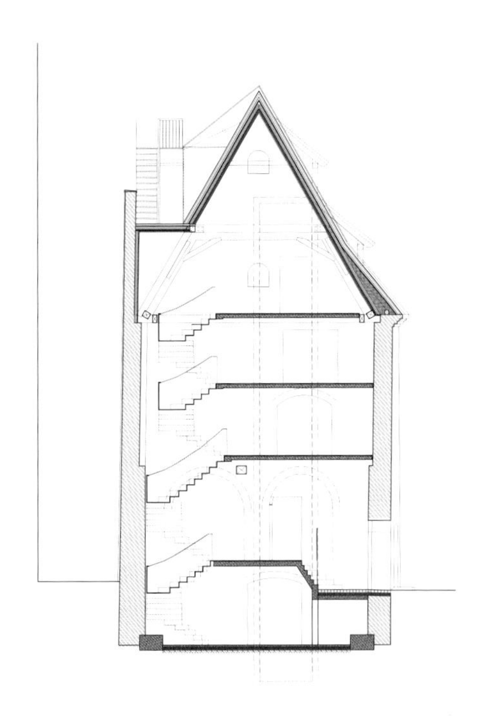

Cross section／横向剖面图

This page: Close-up view of a window. The thickness of the wall is around 85 cm, which has contributed to protect the people in the buiding against cold. Opposite: General view of the building.

本页：窗口特写。这堵墙厚约 85 厘米，能够很好地起到御寒作用。对页：建筑全景。

Sudraba Arhitektūra

"Dome Hotel & SPA", Conversion of 17th Century Apartment Building

Miesnieku Str. 4, Riga 2009

苏德拉巴建筑师事务所

"巨蛋酒店及水疗中心"17世纪公寓楼改造

里加，米斯聂库大街4号 2009

02

201

Dome Hotel & Spa is a contemporary design hotel, merged into the historical substance of a 17th-century apartment building. Historical evidence proves that the basement of the current building has remained from 13th century.

During medieval times (until the mid- 16th century) private property adjoined properties of the bishop and a chapter of priests next to the Riga Cathedral.

During 17th century, until the end of 18th century the residential building had two floors, with the characteristic large hall and mantle chimney (currently serving as the hotel reception). There was a warehouse in the yard, which adjoined Customs on the Pils street and the Dome Square corner. In the end of 18th century the building was supplemented with a cart-house (currently serving as a restaurant) and gateway.

The 3rd floor and the new tiled roof were built in 1870 when a significant amount of rebuilding took place and the building took on its current appearance as well as its layout and aesthetics, which are characteristic of the historicism period. In the early 20th century the apartments were transformed into various workshops.

The architectural design concept is based on retaining balance between preserving and renovating historical elements, contemporary design solutions and satisfying the comfort level characteristic of high class hotels. A sensitive approach towards historical values can be found throughout the building – renovated 17th-century wooden staircase, doors, windows and wall paintings. A new 4-story volume and vertical green wall in the courtyard are the most significant contemporary architecture contributions in this project. Common features of the 15 hotel rooms are maximal use of daylight and a functional, tasteful layout.

Credits and Data

Project title: "Dome Hotel & SPA", Conversion of 17th Century Apartment Building
Client: State Real Estate of Latvia
Location: Miesnieku street 4, Riga, Latvia
Completed: 2009
Architect: Sudraba Arhitektūra
Project team: Reinis Liepins, Ilze Liepina, Ainars Plankajs, Roberts Valdmanis, Ieva Leja, Gundega Duduma, Renate PablakaIG

pp.118–119: View of Riga Cathedral to the southeast from the rooftop terrace of Dome Hotel & Spa. An operable tent roof is installed. Opposite, above: Interior of guest room with well-preserved decorations of 1820s. It used to be a small salon. Opposite, below: View of the hallway on the second floor. On the left, doors with linseed oil coating and restored painting of 1820s are seen. On the right side, there are hallway and window, which were new addition to ensure daylight up to the middle part of the building. Photo by Agnese Zeltina, courtesy of the architect.

第 118-119 页：从巨蛋酒店及水疗中心的屋顶露台眺望东南方向的里加大教堂。屋顶呈可开合的帐篷状。对页，上：客房内景。屋内 19 世纪 20 年代的装饰被很好地保存下来。这里曾经是一间会客厅；对页，下：二楼走廊。左侧可以看到涂有亚麻油的门和修复好的 19 世纪 20 年代的彩画。右侧可见走廊和窗户。这些是为了保证建筑中部的采光而新建的。

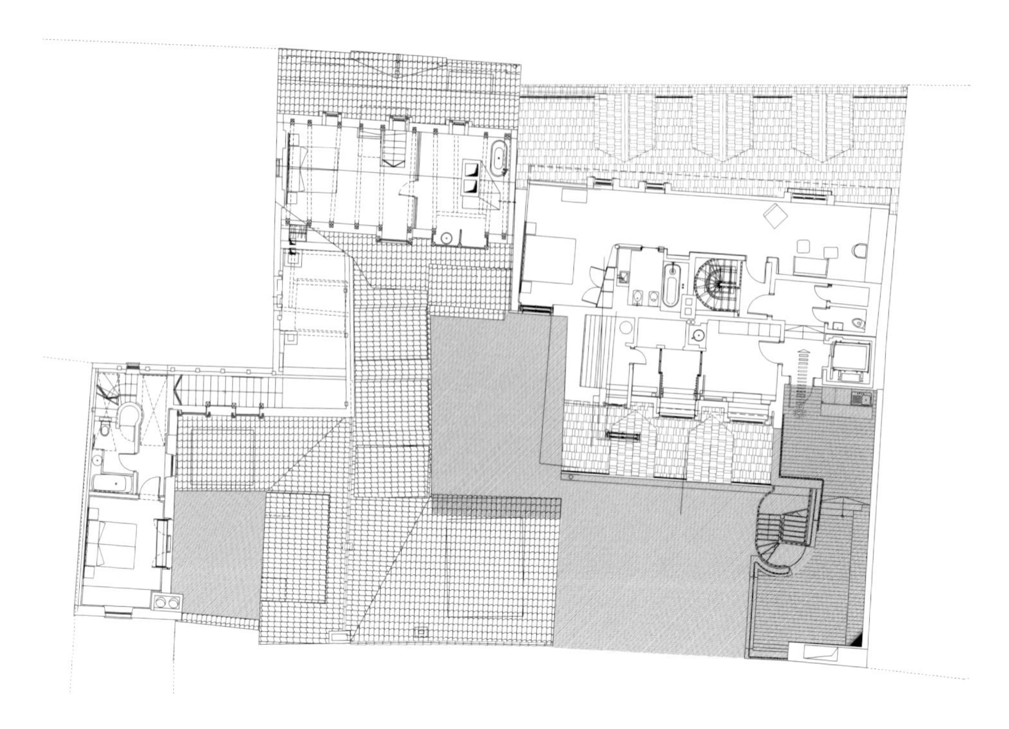

Fifth floor plan／五层平面图

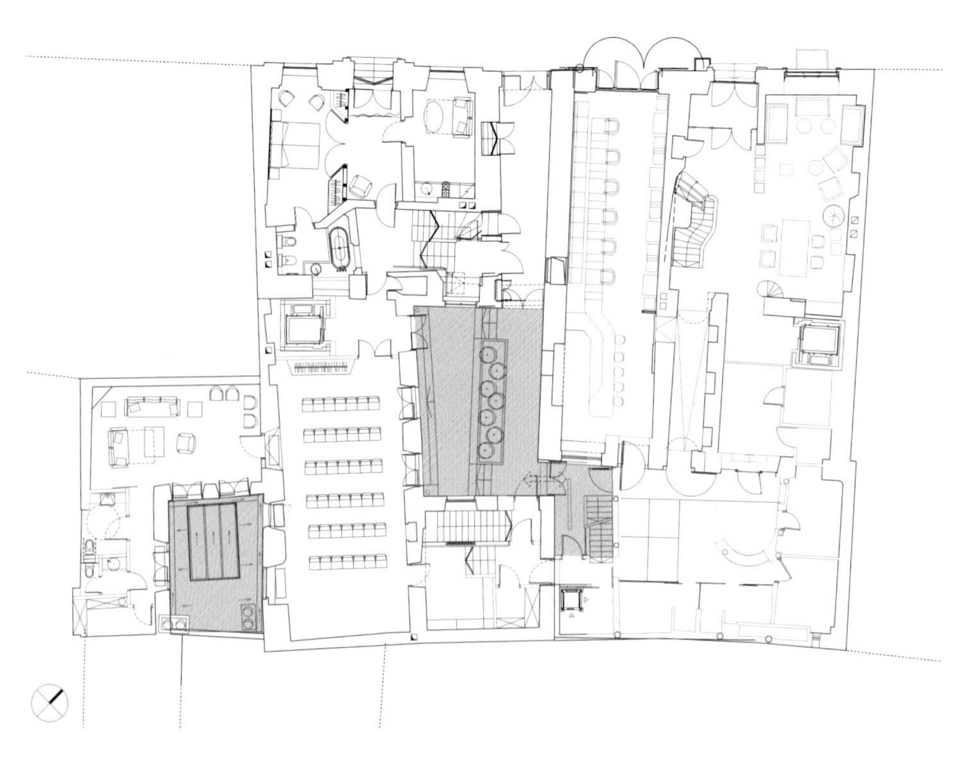

Ground floor plan (scale: 1/300)／一层平面图（比例：1/300）

Section (scale: 1/300)／剖面图（比例：1/300）

巨蛋酒店及水疗中心是一家拥有现代化设计的酒店，同时也融合了17世纪公寓建筑的历史特点。从史料中可以发现，这座建筑的地下室从13世纪开始就已经存在了。

中世纪（直到16世纪中叶），这一私有地产与里加大教堂毗邻，紧挨着主教以及教士的土地。

17世纪到18世纪末，这座住宅楼共有两层，并带有一个极具特色的大厅和壁炉（这里现在用作酒店的接待处）。庭院中建有一座仓库，它与建在皮尔斯大街巨蛋广场拐角处的海关相邻。18世纪末，这座大楼加建了一个车库（现在是餐厅）和大门。

第三层和新修的瓦葺屋顶建于1870年。当时里加的建筑经历了大规模的重建，而这座建筑的外观以及平面布局、美感就形成于这一时期，保留了历史特征。20世纪初，大楼中的公寓房间被改造为各种各样的工作室。

建筑的设计理念是基于在历史元素保留和当代设计之间保持平衡，同时满足高级酒店对舒适性的要求。对历史价值的敏感贯穿了整个改造项目，建筑中仍随处可见翻新的17世纪木楼梯、门、窗以及壁画。而项目中最重要的当代元素就是庭院中新的四层建筑以及垂直绿化墙。15间酒店房间的共同特点是，它们都最大限度地利用了自然光，并具备兼有功能性与品位的布局。

This page: Looking towards outside from the entrance lobby. Delicate hand drawings and the existing wooden structure welcome the guest.
Opposite: Exterior view from the street.

本页：从门厅看向街道。客人一进门，映入眼帘的就是细腻的手绘和原有的木造结构。对页：在街道上看建筑的外观。

DOME HOTEL
&
SPA

Jaunromāns un Ābele

Residential Building on Skārņu Str. 11, Riga

Skārņu Str. 11, Riga 2014

尧恩姆斯与伯勒建筑师事务所

里加斯卡努大街11号集合住宅

里加，斯卡努大街11号 2014

03

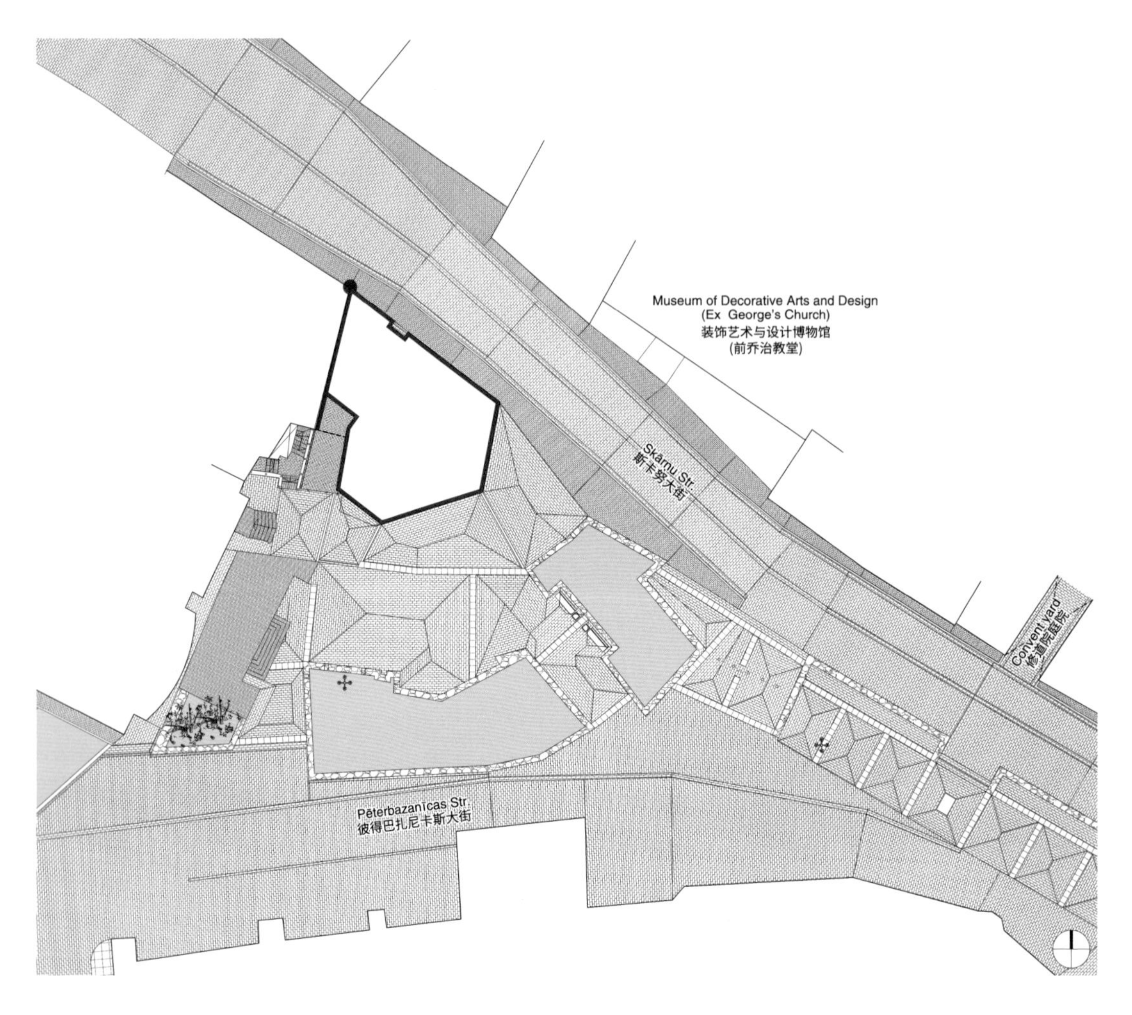

Site plan (scale: 1/600)／总平面图（比例：1/600）

Credits and Data

Project title: Residental Building on Skārņu Str. 11, Riga
Competition: 2006 (1st prize)
Realization: 2013–2014
Architects: Jaunromāns un Ābele (Mārtiņš Jaunromāns, Māra Ābele, Ieva Skadiņa, Jolanta Šaitere, Liene Daņiļēviča)

Creation of this modern seven-level apartment building, located in historical Riga Old Town, was a challenge in both architectural and visual aspects. The design of the building was developed after winning an international architectural competition where participants were given the task of creating a new, high-quality, multi-story apartment block that would be contemporary architectural design yet complement the historic buildings of its location. The solution to establish a new supplement to the historical square that includes some of the most significant buildings in city had to be both visionary and respectful of cultural heritage. The famous neighbors of the new building are St. Peter's Church, a magnificent landmark of gothic architecture, St. John's Church, the oldest church in Riga, and the white dolomite walls of the oldest brick building in town, now home for the Museum of Decorative Arts and Design.

Due to the economic meltdown in Latvia, the project had to be frozen for a while, but it gave architects time to re-think and try out many design ideas and variety of materials. The architects prefer to be closely engaged with all the phases of completing a project, so after construction was finally started, the site was visited daily to review every development. It was decided to use red brick that harmoniously tunes in the ensemble of surrounding buildings, and accent it with a slightly asymmetric chess-board pattern of black metal balconies and windows. After the structural walls were ready, brick samples were placed in window opening to examine the real-life view, in daylight and at night. Brick samples were also tested for temperature, since Latvian winters can be quite rough – sometimes down to -35°C.

The facade was not to remain plain, and after numerous versions, the architects decided to develop idea of using the shadowy silhouette of Bremen musicians – a bronze sculpture from Riga's twin city Bremen. A sculpture depicting a donkey, dog, cat and rooster is located in the square in front of the building. A slightly abstract graphic silhouette of the sculpture, created by the brick pattern, interrupts regularity and gives building an extraordinary and modern dimension. A seamless glue technique was used in the construction of the hanging brick facade, allowing the graphic to be undisturbed by connecting seams.

The outline of the seven-story building replicates the outline of the historic building that once stood on this plot. The courtyard is paved with historic granite cobble stones discovered during architectural research.

Each of the six apartments is designed to have an exclusive 270-degree view, and the top floor apartment has a terrace with a splendid view over Old Town. The building won the highest recognition in the Latvian Architecture Awards competition in 2015, and stands as one of the most interesting modern buildings in the historical Old Town.

pp. 126-127: General view from the south. Photo by Karlis Jaunromans.

第 126-127 页：自南方看到的全景。

这座现代风格的七层公寓楼坐落在里加历史悠久的老城区。它的建造在建筑及视觉方面都是一个挑战。这座公寓楼的设计由一场国际建筑竞赛的角逐而决定。大赛要求参赛者们设计一座全新的高质量多层公寓，既具备当代设计，又是对当地历史建筑的补充。在这样一个遍布城市中最重要历史建筑的广场上打造新建筑，必须既有远见，又尊重文化遗产。这座新建筑毗邻哥特式建筑地标的圣彼得大教堂，里加最古老的教堂圣约翰大教堂，以及拥有白色大理石墙的装饰艺术与设计博物馆（这是城里最古老的砖砌建筑）。

受拉脱维亚经济崩解的影响，这一项目不得不被搁置一段时间，但这也给了建筑师们重新思考设计理念、尝试不同材料的机会。建筑师们更喜欢密切参与一个项目的各个阶段，因此，在项目开始后，他们每天都会到现场查看进展。为了与周围的建筑保持和谐统一，建筑师决定使用红砖，并用呈不对称棋盘式的黑色金属框阳台和窗户来突出效果。在结构墙建好后，砖样被放置在打开的窗口处，用来测试其昼夜的真实效果。此外，由于拉脱维亚冬季的气候条件十分恶劣，有时甚至可以低至-35°C，也对砖样进行了温度测试。

建筑立面并没有保持原状，而是在经过数次修改后，利用不来梅音乐家的影子轮廓来造型。其灵感来源于里加的双子城不来梅城中的一座青铜雕塑。建筑前的广场上有一座刻画有驴子、狗、猫和公鸡的雕塑。建筑师利用砖块来模拟略微抽象的雕塑轮廓，打破了建筑的规律性，赋予其非凡性及现代感。挂砖立面的建造使用了无缝胶技术，使雕塑样式不被接缝打断。

这座七层建筑的轮廓复原了原先矗立在这里的历史建筑的轮廓。中庭则使用在建筑调研中发现的历史悠久的花岗岩鹅卵石铺成。

公寓每一层都可以看到270°的景观，顶层还带有一个露天平台，以供客人眺望里加老城区的壮观景象。这座建筑获得了2015年拉脱维亚建筑竞赛的最高荣誉，现在它已经成为里加历史老城中最有趣的现代建筑之一。

Opposite: Close up view of the facade. The spire in the background is the bell tower of St. Peter's Church. Photo by Martins Jaunromans.

对页：建筑立面。背景中的尖顶建筑是圣彼得大教堂的钟楼。

Sudraba Arhitektūra, MARK arhitekti

Renovation Project of the Castellum of Riga Castle

Pils square 3, Riga 2014–2020

苏德拉巴建筑师事务所，MARK建筑师事务所

里加宫喀斯特勒姆堡改造

里加，比乌斯广场3号 2014-2020

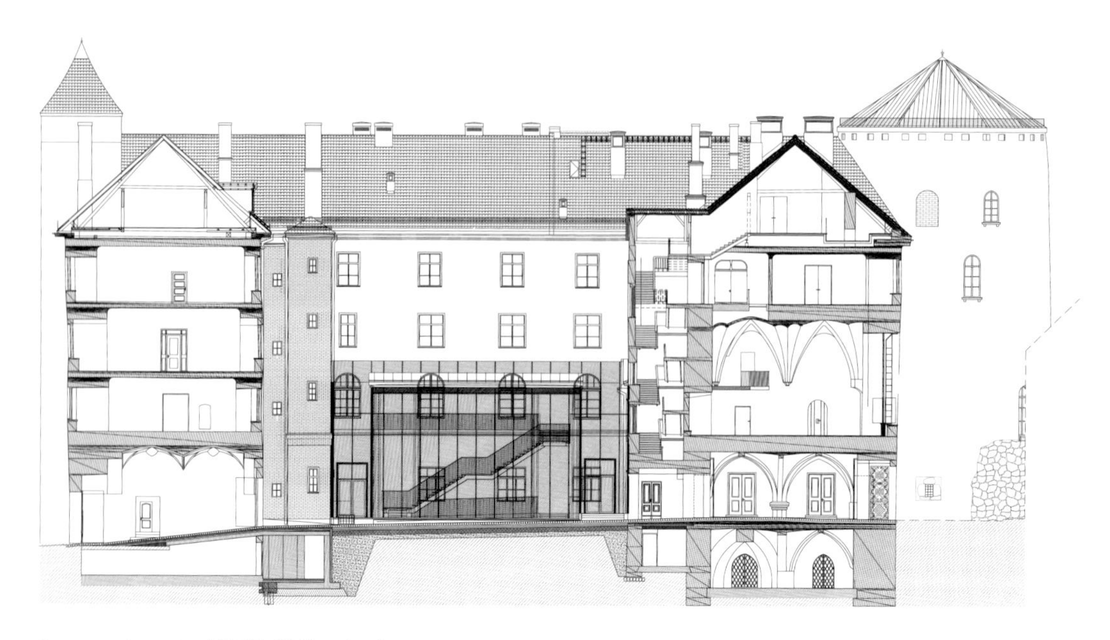

Section (scale: 1/500)／剖面图（比例：1/500）

Riga Castle has stood on the right bank of the River Daugava for more than 700 years and is one of the most significant architectural monuments in the Baltic States. The castle was founded in 1330. Its structure was thoroughly rebuilt between 1497 and 1515. The fortress was continually augmented and reconstructed between the 17th and 19th centuries. Overall there have been around 100 more or less important rebuildings. Currently the northern part of the castle is the official residence of the President of Latvia, but the southern part after the completion of our project will house the Museum of History of Latvia.

Riga Castle is one of the largest medieval castles in Latvia. For many centuries the castle has been a seat of power. Its location in the center of Riga, close to the River Daugava makes the castle even more important geographically, functionally and for representation purposes.

The restoration concept of the Castellum of Riga Castle aims for preservation of historical substance to the maximum extent, adding functionality and technologies required for successful operation of the building in the future. Modifications of the historical part that are envisaged in the design are kept to the minimum and are necessary to reveal the real historical value of the Castellum and to ensure proper functioning of the National History Museum of Latvia.

The Castellum courtyard is part of the urban environment and an extension of Old Riga. The courtyard includes the new two-story extension supplementing the structure of the courtyard arcade. The glass structure is designed as a reference to the former gallery by the eastern wall aiming to reconstruct the courtyard complex. It will contain new staircase, necessary to connect basement, first and second floor. The architecture of the new volume consists of a metal frame construction with concrete footbridges and glass roof.

One of the most important spaces in Riga Castle is chapel located in the 2nd floor. The room is going to be restored to its original appearance by removing the ceiling currently dividing the floors. The concept for interior of the chapel intends to restore an undivided space which is the most significant feature characterizing a medieval castle.

A restaurant is located in the southwest corner of the Castellum, providing a summer terrace which overlooks the River Daugava.

Credits and Data

Project title: Renovation Project of the Castellum of Riga Castle
Location: Pils square 3, Riga, Latvia
Project year: 2015–
Architect: Sudraba Arhitektūra, MARK arhitekti
Project team: Reinis Liepins, Janis Sauka, Ilze Liepina, Ilona Markuse, Liga Rutka, Ainārs Plankajs
Engineering: Engineering bureau Buve un Forma / MARK arhitekti

p. 132, above: Image showing the Castellum's courtyard with the new two-story extension supplementing the structure of the courtyard arcade and containing a new staircase, necessary to connect the basement, 1st and 2nd floors. p. 132, below: Image of the courtyard's basement with the new exhibition hall. On the right, a fortification wall of the tower can be seen. p. 133: Visualization of a bird's eye view of the Riga Castle with the proposed reconstruction and restoration interventions. On the left side, the northern part of the castle is the official residence of the President of Latvia. On the right side, the southern part is the Castellum, currently being restored. Opposite: Image of the interior of the chapel on the 2nd floor. The chapel will be restored to its original appearance by removing the ceiling currently dividing the floors.

第 132 页，上：喀斯特勒姆堡庭院。作为庭院拱廊的补充，院中还有一个新增的两层建筑。它包含有新建的楼梯，用来连接地下室、建筑一层和二层；第 132 页，下：带有新展示厅的庭院地下室。右边可以看到塔楼的加固墙。第 133 页：里加宫改造及翻修效果鸟瞰。左侧是城堡北部用作拉脱维亚总统官邸的部分；右侧是城堡南部，也是正在翻修的喀斯特勒姆堡的所在。对页：二楼小礼拜堂内景。对这一房间的改造旨在恢复其原貌，拆除将空间纵向分割的天花板。

里加宫位于道加瓦河右岸，距今已有700多年的历史，是波罗的海诸国中最重要的纪念性建筑之一。这座宫殿建于1330年，1497年到1515年间其结构被彻底重建，17至19世纪宫殿也在不断扩建、重建。总体而言，宫殿中大约有百余处重要的重建建筑。现在，宫殿北部被作为拉脱维亚总统的官邸；在项目完成后，宫殿南部将成为拉脱维亚历史博物馆。

里加宫是拉脱维亚最大的中世纪城堡之一。几个世纪以来，它一直是权力的象征。此外，它坐落在里加市中心，靠近道加瓦河，这也使得它在地理、功能和象征意义方面更加重要。

里加宫喀斯特勒姆堡的改造旨在最大限度地保留历史遗产，并增设建筑未来成功运营所需的功能和技术。设计中对历史要素的修改被限制在最低限度，以展示喀斯特勒姆堡真正的历史价值，同时也确保它能符合拉脱维亚国家历史博物馆的功能需求。

喀斯特勒姆堡的庭院是城市环境的一部分，也是里加老城的延伸。作为庭院拱廊的补充，院中还有一个新增的两层建筑。它参照东墙处的旧画廊，采用了玻璃结构，旨在重建庭院建筑群；它包含有新建的楼梯，用来连接地下室、建筑一层和二层。这一全新的建筑体由金属框架、混凝土墙和玻璃屋顶组成。

里加宫中最重要的空间之一就是位于二楼的小礼拜堂。对这一房间的改造旨在恢复其原貌，拆除将空间纵向分割的天花板，将其恢复成未经分割的样子。而这也是中世纪城堡最重要的特征。

喀斯特勒姆堡西南角有一家餐厅。夏天，人们可以在餐厅的露台上俯瞰道加瓦河。

Opposite, left: Drawing by Johann Christoph Brotze from the beginning of the 19th century features the third Livonian Order castle in 1515. Image courtesy of the architect. Opposite, right: Image showing the vaulted ceiling of the inner gate of Riga Castle at the end of the 19th century. All image on pp. 132–137 courtesy of the architect. This page,above: Image of the interior of the chapel on the 2nd floor. This page, below: View of the current status of the chapel's vaulted ceiling and the ceiling currently dividing the floors.

对页，左：约翰・克里斯托夫・布罗茨 19 世纪初的绘图，画中描绘的是 1515 年拉特维安政权的第三座城堡；对页，右：19 世纪末，里加宫内城门的拱顶。本页，上：二楼小礼拜堂内景；本页，下：教堂拱顶及分割纵向空间的天花板的现状。

Historical Buildings in Old Town of Riga

里加老城的历史建筑

pp. 138-139: Evening view of Historic Center of Riga from the left bank of the Daugava river.

第 138-139 页：从道尔加瓦河左岸看到的里加历史中心区的夜景。

Swedish Gates and Fortification Wall
Torņu and Aldaru Str. corner
13th century Wall, 1698 Gates.

瑞典门和要塞城墙
托尔尼亚大街和阿尔达鲁大街拐角处
13世纪的城墙，1698年建造的瑞典门

05

Opposite: View towards the Swedish Gate at the corner of Torņu and Aldaru streets. The Swedish Gates were built in the existent fortification wall to provide access to barracks outside the city walls at the end of the 17th century. This fragment of the wall is the oldest remaining part of the Old Town fortifications as the remainder was progressively knocked down. This page: View of the typical narrow, winding Krāmu street in the Historic Center of Riga. At the back, spire of Rīgas Cathedral is seen. The corner of Krāmu and Jauniela streets is marked by an Art Nouveau building, crowned with a small tower that resembles the spires of the churches characteristic of the medieval Old Town.

对页：自托尔尼亚大街和阿尔达鲁大街的拐角处看向瑞典门。瑞典门建造于 17 世纪末，位于已有的要塞城墙内，是通往城外兵营的出入口。由于城墙逐渐被破坏，现在残存的遗迹是旧城要塞城墙中最古老的部分。本页：克拉姆大街是里加历史中心区典型的狭窄道路。背景可见里加大教堂的尖塔。克拉姆街和雅尼耶拉街的拐角处建有一座新艺术风格的建筑，建筑顶部有一座小塔楼，与中世纪里加老城教堂的尖塔类似。

Apartments and Retail Premises
Alfrēds Ašenkampfs, Maksis Šervinskis
Audēju Str. 7
1899

公寓和商业建筑
阿尔弗雷德斯·阿什坎普夫斯，马克斯·谢林斯基
奥德朱大街7号
1899年

06

pp.142-143: View of Dome square. St. John's Church (13th–17th century) in the middle and at the right St. Peter's Church at Skārņu (13th–18th century, reconstruction in 1968–1977) are seen. St. John's Church marks the location of the former residence of Bishop Albert of Riga (13th century). This page: Street view from the southeast. The building in front is originally an apartment and retail premises building and is one of the early examples of Art Nouveau in Riga and is also a large-scale reconstruction conceived in 1899 as a transformation of an older building. Opposite: View of the main nave of St. Peter's Church. The interior is an example of the Gothic brick structures of Northern Europe.

第 142-143 页：巨蛋广场。中间是圣约翰教堂（13–17 世纪），右侧是位于斯科鲁街的圣彼得教堂（13-18 世纪，1968-1977 年重建）。圣约翰教堂建在里加阿尔伯特主教（13 世纪）故居的遗址上。本页：自东南方看街景。前方的建筑原本是一座公寓及商业建筑，也是里加新艺术风格建筑的早期范例之一，于 1899 年经历了大规模改造。对页：圣彼得教堂主厅。内部是典型的北欧哥特式砖结构。

St. Peter's Church
Skārņu Str. 19, Kaļķu Str. 2 and Grēcinieku Str.11
13–19th century, 1968–1977 (reconstruction)

圣彼得教堂
斯科鲁街19号，开丘街2号及格列蒂尼克街11号
13-19世纪，1968-1977年重建

07

Opposite: View of the roofscape of the Old Town of Riga from Riga Castle. At the back, the spires of Riga Cathedral and St. Peter's Church. Front, Our Lady of Sorrows Church, built in the 18th–19th century, is seen. This page: View of the roofscape of the Old Town of Riga from St. Peter's Church. In the middle, structures of Riga Central Market's four pavilions that originally were used for the German Zeppelin hangars, Latvia during World War I. At the back, Riga Radio and TV Tower (1979–1989).

对页：自里加宫看到的里加老城区的屋顶景观。在背景中可以看到里加大教堂以及圣彼得大教堂的尖塔。眼前的是建于 18–19 世纪的圣母受难教堂。本页：自圣彼得教堂看到的里加老城区的屋顶景观。视野中央是里加中央市场的四个分馆，它们最初是拉脱维亚在“一战”时用来停放德国齐柏林飞艇的飞机库。在背景中可以看到里加广播电视塔（1979–1989）。

Riga Central Market

Pāvils Dreijmanis, Pāvils Pavlovs, Vasilijs Isajevs, Georgs Tolstojs
Nēgu Str. 7
1924–1930

里加中心市场
帕韦尔斯·[illegible]马尼斯，帕韦尔斯·帕罗斯乔治·托尔斯托夫斯
耐谷大街7号
1924–1930

08

pp.148-149: Interior view of the meat pavilion at Riga Central Market, showcasing the metal structures of the former Zeppelin hangars.

第 148-149 页：里加中央市场中肉类市场的内景，展示了前齐柏林飞机库的金属构架。

Essay:

Background of Architecture in Riga, Latvia

Jānis Lejnieks

论文：

里加的建筑背景

亚尼斯·雷伊尼克斯

States come and go; their borders change while towns and cities are eternal. Warriors of the Crusades instigated by the Pope arrived to convert pagan tribes to Christianity. Bishop Albert supported German trader interests in this region and in 1201 established Riga as a Livonian capital on the banks of the Daugava River. There were still 700 years to wait until the establishment of an independent Latvian state, and the Latvian nation had not yet formed out of separate tribes – that happened only in the 19th century. Presently almost half of all Latvia's inhabitants live in Riga, and it is a more popular brand than Latvia.

German conquerors entered in the 12th century bringing not only the Christian faith but also a new building technology, namely masonry walls that outmatched the wooden log castles of local tribes, easy to set on fire. German castles were, for their part, destroyed by a Russian army that used even more advanced war technologies. When coming under the rule of Russia in the 18th century, in Latvia, as a province, classic styles flourished in the architecture of manors and churches designed and built by local building masters; top designs were however created by guest architects. The most powerful vertical structure in the Latvian architectural landscape is St. Peter's church in Riga built by Rupert Bindenschu (1645–1698), a master builder from Strasbourg. With a spire of 121 m, it was the highest wooden building in Europe of the time. This vertical accent is counterbalanced by an equally expressive horizontal structure, Rundā le Palace (fig 1), designed and built by Bartolomeo Francesco Rastrelli (1700–1771), an Paris-born architect working in the Russian tzar's court.

The first local architect from Riga, Christoph Haberland (1750–1803), additionally studied in Berlin and afterwards created noble classic buildings in his fatherland. At the start of the 19th century, the creative genius of the Latvian people could only grow in the form of national-style buildings, i.e. improvements in log building technique. The best examples of this technique have been collected in Riga, in the Latvian Ethnographic Open-Air Museum established after proclamation of an independent Latvian state in 1918 (fig 2). However, such Latvian master builders as Mārcis Sārums (1799–1895) fulfilled orders from German landlords and church parishes and acquired skills in practical work.

Latvia, as a Western province of Russia, experienced rapid growth in the 19th century, and newbuilds were designed by local architects, with most of them academically educated in Riga Polytechnical Institute's Faculty of Architecture founded in 1869. Nowadays, it is the Faculty of Architecture and Urban Planning within Riga Technical University; also, one may study to be an architect at the Faculty of Architecture and Design at Riga International School of Economic and Business Administration. However, the first professional Latvian architect, Jānis Fr drihs Baumanis (1834–1891), studied in Berlin's Bauakademie and received his diploma from the St. Petersburg Imperial Academy of Arts. Like his Baltic-German colleagues, J.F. Baumanis followed the principles of eclecticism that were dominant in Western Europe and designed many buildings at Riga Boulevard Circle, formed after dismantling medieval fortifications in the mid-19th century (fig 3). An expressive testimony of military architecture

政权交迭，国界移动，但城镇永恒。十字军在教皇的煽动下抵达这里，试图让异教徒改信基督教。阿尔伯特主教支持德国商人在当地的贸易活动，1201年在道加瓦河畔建立了里加，将其作为利沃尼亚的首都。不过此时距离这片土地独立成为拉脱维亚，还有700年的时间；拉脱维亚一统分散的部落，也是到19世纪才发生的事情。现在，近半拉脱维亚的居民都生活在里加，那是比拉脱维亚更受欢迎的招牌。

12世纪德国的征服者不仅带来了基督信仰，还有新的建造技术——砖砌墙，这比当地易遭火灾的木质城堡更胜一筹。然而，即使是这样的德国城堡，在面对俄罗斯军队使用的更先进的战争武器时，也难逃一劫。18世纪，拉脱维亚作为行省臣服于俄罗斯帝国的统治。此时古典风格大行其道，当地工匠设计、建造的庄园和教堂也多为此类，但顶级的方案仍出自外籍建筑师之手。拉脱维亚建筑景观中，最有力量的垂直结构，是法国斯特拉斯堡的建筑大师鲁珀特·宾得丘(1645-1698)的里加圣彼得教堂。它是当时欧洲最高的木结构建筑，塔楼尖顶高达121米。相对地，还有一座同样富有表现力的水平结构建筑——伦达尔宫(图1)，设计者为巴黎出生的建筑师弗朗切斯科·巴尔托洛梅奥·拉斯特雷利(1700-1771)，当时他正为沙俄皇室效劳。

而里加的第一位本土建筑师，是克里斯托夫·哈伯兰(1750-1803)，他曾在柏林学习，之后回到故土设计高贵的古典主义建筑。尽管身怀创造力，这位拉脱维亚天才在19世纪初也仅仅是发展了民族特色的建筑形式，例如改良原木建造技艺。该技艺的最佳应用范例已被收藏到1918年独立宣言发表后创建的拉脱维亚民族史露天博物馆(图2)。至于马尔西斯·斯鲁姆斯(1799-1895)之类的拉脱维亚建造大师，则是在完成德国地主和教区的要求中掌握了实践技巧。

19世纪，作为俄国西部省份的拉脱维亚快速发展，新的建筑设计任务都交给了本地建筑师。这些建筑师中的大部分都曾就读于1869年创办的里加理工大学建筑系。放在今天，就是里加科技大学建筑与城市规划系；顺便一提，里加国际经济与工商管理学院也有建筑设计系。里加第一位职业建筑师是亚尼斯·弗里德里希斯·鲍马尼斯(1834-1891)，他先赴柏林建筑学院学习，之后从圣彼得堡皇家美术学院拿到学位。与他那些波罗的海德裔的同事一样，鲍马尼斯遵循西欧主流的折衷主义原则，在19世纪中期取代中世纪防御工事的里加环形大道沿线，设计了许多建筑(图3)。陶格夫匹尔斯城堡是现存特色鲜明的军事建筑，有着出色的防御结构，整体占地150公顷，也是东欧唯一一座几乎从19世纪就完整保存下来的碉堡式防御工事。

里加逐渐发展为俄罗斯帝国的第二大港口城市和大型工业中心，到1914年之前，居民人口超过了50万。经济快速增长的同时，新艺术的风潮从南方和北方一同涌入拉脱维亚。在俄国新贵的期待下，德国新艺术的装饰性在阿尔伯特街得到了展现(图4)；但此时，下一代的拉脱维亚建筑师已经把注意力转移到了其他案例上。芬兰建筑师的民族浪漫主义风格的建筑，以及拉脱维亚的民族建筑，都对他们产生了很大影响，启发他们创造属于自己的杰作。艾森斯·劳贝(1880-1967)和亚历山大斯·瓦纳格斯(1873-1919)就是在20世纪初到访过赫尔辛基之后，创造出了所谓的北欧风格建筑，并以新艺术的名义构成了里加中心最具价值的部分。

“一战”之后，现代建筑仍以缓慢的速度进入拉脱维亚，波罗的海德裔建筑公司Karr&Baetge及其建筑工程师就是在包豪斯风格的启发下进行设计。但也有例外，例如建筑师阿尔弗雷兹·格林博格(1893-1940)设计的里加学

is preserved in Daugavpils Fortress, an outstanding fortification structure with a total area of 150 ha; it is the only 19th century bastion-type fortification in Eastern Europe preserved in an almost untransformed state.

Riga developed as the empire's second largest port city and a large industrial center, with the number of inhabitants exceeding half million before 1914. Rapid economic growth coincided with art nouveau's expansion entering Latvia from both south and north. German art nouveau's decorativeness, according to the wish of the Russian nouveau riche, was expressed in the curious Alberta street (fig 4) development while the next generation of Latvian architects focused on other examples. A leading impact was made by buildings designed in the style of national romanticism by Finnish architects and models of Latvian national building that inspired Latvian architects to create their best works. After visiting Helsinki at the start of the 20th century, Eižens Laube (1880–1967) and Aleksandrs Vanags (1873–1919) created the so-called Nordic style buildings that form the most valuable part of Riga center under the name of art nouveau.

After World War I modern architecture was slow to enter Latvia. Baltic-German architectural company Karr&Baetge and building engineers created Bauhaus-inspired designs. School buildings in Riga designed by architect Alfrēds Grīnbergs (1893–1940) and Latvian Red Cross buildings by Aleksandrs Klinklāvs (1899–1982) were an exception to that. In the citadel of architectural theoretical thought, the Faculty of Architecture at the University of Latvia, there was no consensus on supporting modern architecture because of three differently-minded workshops. Ernests Štā lbergs (1883–1958), head of Workshop "C", did work under the inspiration of functionalism, yet was not a propagator of the style. Quite possibly, this was one of the obstacles to the expansion of contemporary architecture in Latvia; this is supported by the fact that works by F.L. Wright, Le Corbusier and Mies van der Rohe were not known to the wider public in Latvia up to as late as the 1960s. Workshop "B" was led by Pauls Kundziņ š (1888–1983) (fig 5), an active student of people's building style, who designed functionalistic buildings embellished with ornaments. "A" was led by Laube who had come to favor American 19th century classicism. Latvian authoritarian state paved the way to a totalitarian architecture gaining way after the incorporation of Latvia into USSR in 1940. A striking example to this is the Palace of Justice (fig 6) designed by architect Fridrihs Skujiņš (1890–1957) who emigrated to Germany (the building was finished after war by an architect from Russia who retained its original style). Although the regimes had changed, the monumental governmental building style that had satisfied the independent state continued to meet

交建筑，亚历山大·克林克拉夫斯（1899-1982）设计的红十字会大楼。在建筑理论的大本营——拉脱维亚大学建筑系，三个工作坊各持己见，对现代建筑始终没有达成共识。工作坊“C”的负责人埃内斯茨·斯托尔伯格斯（1883-1958）尽管受到功能主义的启发，却没有将这种风格发扬光大。这很可能是阻碍当代建筑在拉脱维亚扩散的因素之一，毕竟直到20世纪60年代，弗兰克·赖特、勒·柯布西耶、密斯·凡·德·罗等著名建筑师的作品才被更多的拉脱维亚民众所知。工作坊“B”由帕乌尔斯·昆津修（1888-1983）（图5）领头，这是一位积极应用大众建筑风格的学生，设计了一些带有装饰的功能主义建筑。工作坊“A”的主持者劳贝，则倾向于美国19世纪流行的古典主义。1940年，拉脱维亚并入苏联之后，专制政权为极权主义建筑铺平了道路，其中一个引人注目的例子，便是移民德国的建筑师弗里德里希斯·斯库伊修（1890-1957）设计的拉脱维亚最高法院——司法宫（图6，建筑在战后的收尾工作由一位俄罗斯建筑师完成，原本的样式得到保留）。虽然政权不断更迭，但各时期的统治者都准许建造这种适用独立国家的纪念性政府建筑。

苏联时期现代主义建筑最突出的例子是莫德里斯·格尔吉斯（1929-2009）在帕巴兹地区建造的夏日别墅（图7），这是拉脱维亚有史以来登记在册的最小建筑。通常来说，乡村的传统农场并入周围的农业小镇，较大的城镇模仿英式、北欧的住宅规划模式。然而，苏联的经济无力支撑这样的转型，到20世纪70年代，苏联的现代主义（与西方的国际风格十分类似）透支了社会给予的无条件信任，也未能带来承诺过的幸福生活。虽然此时政治权力和建筑工业的中心在莫斯科，但里加的精英建筑师选择了另一条通往现代主义的道路，以一种后现代主义的形式美化过去，也迎合了社会的怀旧情结。1979年，年轻的建筑师在拉脱维亚建筑协会举办了一场名为“地方建筑何去何从”的重要展览，聚焦苏维埃政权想要一改全貌的乡土建筑。20世纪80年代培养了老一辈建筑师，他们在20世纪90年代，苏联现代主义建筑因偶像崇拜废除而被破坏、复制的历史建筑取而代之时，保持了保守态度。直到1997年里加历史中心被联合国教科文组织列为世界遗产名录，这一过程才宣告结束。如今，里加的代表仍然是巴洛克式教堂、环形大道、“新艺术”街区以及在21世纪士绅化进程中复兴的工人阶级的郊区木屋。

自从20世纪30年代以来，里加始终梦想着建立一个新的中心，也在道加瓦河左岸寻找合适的地点。1991年拉脱维亚恢复独立后，本国出生的建筑师古纳·伯克茨接到委托，设计拉脱维亚国家图书馆（见170-171页）。该项目已竣工，不过现在老城区以北（仍在道加瓦河右岸）的斯堪斯特区有望发展为商业中心。20世纪90年代，随着国界打开，新的材料和思想进入拉脱维亚，多元的氛围也给恢复过去的建筑表现方式提供了机会。修复传统在拉脱维亚得到了很好的传承和发展，使得围绕拉脱维亚建筑特征，本期客座主编伊尔泽·帕克罗内概括为“拉脱维亚性”）的讨论不断升温。基于“二战”给里加老城带来的破坏，人们开始思考，怎样的建筑才有权参与老城的复兴。事实上，第

5

6

with the approval of the occupational authorities.

The most astounding example of soviet-time modernism is Modris Ģelzis' (1929–2009) summer house in Pabaži (fig 7), the smallest listed building of any period in Latvia. In the rural landscape, the traditional farmstead was being joined by small agrarian towns while in larger towns and cities the progressive English and Nordic dwelling area planning model was taken as an example. Due to the soviet economic inability their adaption was however not successful, and in 1970s it became clear that soviet modernism, very much like the Western International Style, had used up the society-issued blank check without providing the promised happiness. The center of political power and the building industry was in Moscow, yet the cultural elite architects in Riga chose an alternative to modernism, a romanticized past in the form of post-modernism, that satisfied societal nostalgia. The young architects' exhibition of 1979 titled "What will be the fate of regional architecture?" at Latvian Architects Association was of importance and focused on vernacular architecture that the soviet regime had condemned to replace. The period of 1980s nurtured the present older generation who expressed consistently conservative attitudes later in 1990s when soviet modernist buildings were torn down in the spirit of iconoclasm and copies of historic buildings set up instead. The process was stopped by the inclusion of Riga's historic center in the UNESCO World Heritage List in 1997. Riga is still being characterized by baroque church silhouettes, Boulevard Circle, art nouveau quarters and wooden buildings in working class suburbs that have returned to life through a gentrification process in the 21st century.

Riga has been dreaming of a new center since 1930s, and has looked for a respective space on the Daugava's left bank, too. After restoration of the independent Latvian state in 1991 Riga commissioned the Latvian-born architect Gunā rs Birkerts to design the Latvian National Library (See pp. 170–171). This project has been recently realized, but it is rather the Skanste area to the north of the Old Town that is developing as a central business district. The 1990s were a time when borders opened and new materials and ideas arrived, thus providing a chance to restore architectural expression in a pluralistic atmosphere. In Latvia restoration traditions are well developed, and this has heated up speculation on Latvian architectural identity or, what the journal's guest editor Ilze Paklone characterizes as 'Latvianess'. Damage brought by World War II in Vecrīga has been a basis for argument on what architecture has the right to participate in old town's revival. As a matter of fact, the war is not over: discussions on the development of the city's oldest square are going on, along with museification counterbalanced by innovative quests for form (Andris Kronbergs et al.).

论并没有结束：关于如何开发城市中最古老的广场、博物馆化及创新模式的探讨尚未分出高下（安德里斯·克朗博格等）。

Fig 1: Rundāle Palace.
Fig 2: Threshing barn. Latvian Ethnographic Open-Air Museum. Riga. Courtesy of the author.
Fig 3: Aerial photo of Riga. Photo by Juris Kalniņš, courtesy of the author.
Fig 4: View of Alberta street.
Fig 5: Apartment building at Baznicas 8 by Pauls Kundziņš. Courtesy of Latvian Museum of Architecture.
Fig 6: The Palace of Justice by architect Fridrihs Skujiņš. Courtesy of Latvian Museum of Architecture.
Fig 7: Summer house in Pabaži by Modris Ģelzis. Courtesy of Latvian Museum of Architecture.

图 1：伦达尔宫。
图 2：脱粒谷仓，里加拉脱维亚民族史露天博物馆。
图 3：里加鸟瞰图。
图 4：阿尔伯特街。
图 5：巴兹尼卡斯 8 号的公寓楼，帕乌尔斯·昆津修设计。
图 6：弗里德里希斯·斯库伊修设计的司法宫。
图 7：莫德里斯·格尔吉斯设计的帕巴兹夏日别墅。

Jānis Lejnieks, born in 1951, is a member of the Scientific Council of the State Inspection for Heritage Protection, the economic development manager of the Riga City Council Eastern Executive Board, the founding editor-in-chief of a bi-monthly magazine *Latvijas Architektūra* (Latvian Architecture), and project manager of Latvian Town Planning Institute. He received the architect qualification in 1975 and Dr. Arch. in 1994 from Riga Technical University. He has taught history of architecture and architectural design at Riga Technical University, Faculty of Architecture as associate professor in 1994-1999 and as guest lecturer in 1993-1999.

亚尼斯·雷伊尼克斯，生于1951年，国家遗产保护审查科学委员会成员，里加市议会东部执行委员会经济发展部主任，双月刊杂志《拉脱维亚建筑》创刊主编，拉脱维亚城镇规划所的项目经理。他曾就读于里加科技大学，1975年获得建筑师资格证，1994年攻下建筑博士学位。他在里加科技大学建筑系教授建筑史和建筑设计，1994-1999年担任助理教授，1993-1999年为客座讲师。

NRJA

Unwritten – Exposition of Latvia

The 14th International Architecture Exhibition – la Biennale di Venezia 2014

NRJA

拉脱维亚展："Unwritten"

第14届国际建筑展——威尼斯双年展 2014

The exposition with the project title Unwritten highlights issues regarding the perception, research and conservation of Latvian post-War modernist architecture. Unwritten chronicles, in fact, nonexistent research on it.

The exposition is based on the assertion that There is (no) modernism in Latvia which is equally correct as proof and denial of the modernist architecture. There is no acknowledged research and evaluation of post-War modernist architecture in Latvia. The situation is complicated regarding the evaluation of modernist architecture. On one hand, people dislike anything that occurred during the period of Soviet occupation and the influence; on the other, there is a wave of uncritical nostalgia for the country's youth and childhood, as well as a superficial hipster joy at the exotic Soviet heritage.

A number of these buildings have already reached the "monument" age. There is no single evaluation of their importance in the context of the Latvian architectural heritage. We continue to lose them through various strategies of absorption – rebuilding or demolishing. If we continue this way and at the current rate, we will not leave future generations the opportunity to see and evaluate the design and buildings of this period.

Taking into account that both the local and international community and architectural professionals are interested in the Latvian post-war modernist architecture, the UNWRITTEN curators (in October, 2014) asked the State Inspection for Heritage Protection to include 11 buildings in the list of protected cultural heritage monuments. For example: from the 60s of the 20th century one can find unique curves of a thin concrete roof at the gas station in Ogre. It has been used as prototype for other gas stations in some Soviet republics. Sadly, after collapse of Soviet Union in the early 90s, most of the buildings from that time were considered as not valuable and were demolished or abandoned - also gas stations. Now this is the last one surviving piece of that type. It was built in 1965 and has been as gas station till now and hopefully will survive in the future.

Slowly but surely Unwritten has initiated the process of evaluating Modernist Architecture – based on the gathered materials and application at the State Inspection for Heritage Protection. After serious research by the Heritage Protection specialists the gas station in Ogre along the administrative building in Āgenskalns (Riga), the TV tower in Zaķusala (Riga), and the restaurant "Sēnīte" have already been accepted as protected cultural heritage monuments. So, it is official – these are state the protected examples of Latvian modernist architecture.

Amongst 11 candidates there was also monument at site of a concentration camp in Salaspils which is already in heritage list as monument of historical importance. Now it waits to be recognized also as architectural monument.

Today something has really changed – architects society and society in general with stronger confidence can get involved in discussion about conservation and protection of Latvian post-War modernist architecture. Encouraged by successful examples and positive changes in attitude towards this part of architectural heritage some other organizations and local authorities have asked the State Inspection for

展览"Unwritten"（译为：未写完的）展出了拉脱维亚战后的现代主义建筑，着重关注对它的认知、研究和保护。事实上，关于这段无记载的历史，甚至不存在相关的研究。

"拉脱维亚（是否）存在现代主义"，是策展的基本前提，它也证明（或否认）了现代主义建筑的存在。然而在拉脱维亚，对于战后的现代主义建筑，并没有公认的研究或一致的评价。评价时往往需要考虑复杂的背景：一方面，人们不喜欢苏联占领时期的一切事物，以及那种影响；另一方面，存在一股不加批判的怀旧风潮，人们能从具有异国特色的苏联遗产中获得粗浅的愉悦感受。

其中的部分建筑已经存在很久，足以被视为"纪念物"。但在拉脱维亚建筑遗产的语境中，它们的重要性从未得到评定，而且我们正在通过各种自我消化的策略，包括重建或拆除，任由其消失。以目前的速度来看，如果继续这样做，我们的下一代将没有机会一睹这些设计和建筑，更无法做出评价。

考虑到拉脱维亚战后现代主义建筑引起了国内外建筑界的关注，"Unwritten"策展团队（2014年10月）向国家遗产保护审查委员会提出申请，将11处建筑列为受保护的文化遗产。其中之一是位于里加东南的奥格雷的一座加油站，你会发现它有着单薄的混凝土屋顶，曲线颇为独特；这种式样曾作为原型，自20世纪60年代起被应用于苏联其他一些加油站。遗憾的是，90年代初苏联解体后，旧时代的大部分建筑被认为没有了价值，故而被拆除或废弃，加油站也不例外。这是该类型建筑最后一个鲜活的样本，建于1965年，以加油站的形态使用至今，希望在未来仍能继续被保留。

或许进展缓慢，但可以确定的是，"Unwritten"通过搜集材料并向国家遗产保护审查委员会提交申请，已经开启了现代主义建筑的评估进程。经过审查专家的慎重研究决定，奥格雷的加油站与阿根斯卡恩斯（里加）的行政大楼、扎库萨拉（里加）的电视塔、餐厅"Senite"一同被列为文化遗产。所以，现在它们名正言顺地成为受国家保护的拉脱维亚现代主义建筑范例。

11个候选建筑中，还有一处位于萨拉斯皮尔斯的集中营。它已经作为具有历史意义的纪念碑跻身遗产之列，但仍在等待被评定为建筑遗产。

时代在变，无论建筑界还是普遍意义上的社会，都有了更大的信心，来参与讨论拉脱维亚战后现代主义建筑的保存和保护。成功的申遗案例、意识的积极转变，都鼓舞了其他一些组织和地方当局，他们开始请求国家遗产保护审查委员会考虑更多的候选对象。与此同时，公众活动也在进行。例如，以建筑师扎格·盖雷为首的团体就抗议拉脱维亚被占时期博物馆的扩建，因为这间博物馆正是里加老城中心现代主义建筑的最佳典范之一。抗议颇有成效，扩建项目的设计和审批都已中止。

"Unwritten"由三部分组成：一个网站，一处在阿森纳美术馆的艺术装置，一部名录。

威尼斯的拉脱维亚展"Unwritten"部分已在线上虚拟呈现，可以访问www.facebook.com/Unwrittenlv。这是一个开放的数据网站，介绍了战后拉脱维亚现代主义建筑的相关作品和建筑大师。任何人都可以增加拉脱维亚建筑的条目，只要它能够被视为现代主义。由此，"Unwritten"将成为有史以来最大的战后拉脱维亚现代主义建筑的数据库，也将激发未来的研究。

"Unwritten"的未来是开放的。我们不是研究者，我们只是作为公众、调解人和消息提供者，服务于拉脱维亚

Ieritage Protection to include new objects into
he list. There are also some public activities.
or example:a group led by architect Zaiga Gaile
merged with protests against the project for an
xtension of museum of Occupation – one of the
inest modernist architecture examples right at
he heart or Riga Old city center. It went well,
lesign and approving process of extension has
topped.

UNWRITTEN consists of three parts: a website,
n installation at Arsenale and a printed catalogue.

Part of the Latvian exhibition – Unwritten – is
irtual, and can be accessed at www.facebook.com/
nwrittenlv. This is an open social information
ite about architecture and masters who could be
onsidered part of post-War Latvian modernist
rchitecture. Everyone is invited to add examples
f Latvian architecture that could be considered
nodernist. Unwritten will become the largest ever
ase for post-War Latvian modernist architecture
latabase. It will become an initiative for future
esearch.

The future of Unwritten is open. We are not
esearchers, we only serve as a public figure, a
noderator, a news provider about modernist
rchitecture heritage in Latvia. We inform about
ther researches, we follow international press
bout modernist heritage and spread it to local
eople. If someone wants to write a research, we
re ready to help and to get involved in the process.
s there or not modernist architecture in Latvia?
The joy of answering this question we would like
o leave to the others: readers, state institutions or
esearchers in, hopefully, near future.

的现代主义建筑遗产。我们也关注其他研究，跟进有关建筑遗产的国际动态，并将这些传达给拉脱维亚的大众。如果有人打算撰写研究报告，我们也愿意给予帮助并参与其中。至于拉脱维亚究竟是否存在现代主义建筑？我们想把解谜的乐趣留给其他人，例如读者、政府机构、或是有望在不远的将来出现的研究者。

Oppsition, from top: Gas station in Ogre, 1965; Administrative building for Pārdaugava district council in Riga, 1975; TV and Radio tower of Riga in Zaķusala , 1986. This page, from top: Cafe and Restaurant "Sēnīte", 1967; Salaspils Memorial Complex, 1966. All photos on pp. 114–115 courtesy of NRJA.

对页，从上到下：奥格雷的加油站（1965）；里加帕道加瓦区议会的行政大楼（1975）；扎库萨拉的里加广播电视塔（1986）。本页，从上到下：咖啡馆餐厅"Senite"（1967）；萨拉斯皮尔斯纪念建筑群（1966）。

Left Bank of the Daugava River

道加瓦河左岸

The left bank of the river Daugava has the surprising character of being in-between urban and rural spatial organization principles. The streets are laid out straight and buildings set along the streets as in the city center, yet hiding silent and vast gardens at the backyards as in the rural areas. The island of Ķīpsala in the Daugava, just across from the historical center on the right bank of the river, was built up gradually with small wooden residential buildings, starting from the 17th century, and is now a remarkable ensemble of wooden architecture that is developing into a high-quality residential district. The left bank of the Daugava has been the subject of many debates and spatial development proposals, as the territory still has vacant areas to receive new structures of high urban intensity and density.

道加瓦河左岸的空间组织原则介于城市和乡村之间，独具特色。与市中心相似，这里的街道规划得笔直，沿街建筑物鳞次栉比；不过，建筑后面往往隐藏着静谧的大庭院，这又与乡村地区类似。道加瓦河中的普萨拉岛与右岸的历史区遥遥相对。从 17 世纪开始，这座岛上逐渐建成了小型木构住宅，现已形成了木构建筑群，正发展成为高级住宅区。道加瓦河左岸还存在有可以容纳高强度、高密度城市结构的空地，因此它也成为众多空间开发讨论及提案的中心议题。

03
05
04
06
07
01
02
New town
新城
Old town
老城
Left bank
左岸
Daugava River
道加瓦河

Riga Concert Hall
SZK and Partners
AB Dambis, Riga
2006 (pp. 164–169)

Latvian National Library
Gunnār Birkerts
Mūkusalas Str. 3, Zemgales Neighbourhood, Riga
2014 (pp. 170–171)

Gypsum Factory Renovation
Zaiga Gaile Office
Balasta dambis 70, Ķīpsala neighbourhood, Riga
2004–2015 (pp. 172–177)

Žanis Lipke Memorial
Zaiga Gaile Office
Mazā Balasta Str. 8, Ķīpsala neighbourhood, Riga
2012 (pp. 178–183)

Photo Credit
02: Courtesy of Jānis Lejnieks / 04:Photo by Ansis Starks, courtesy of the architect /06; Photo by A. Sturmanis, courtesy of the architect /07:Courtesy of NRJA

The Collection of Wooden House in Ķīpsala
Zaiga Gaile Office
Balasta dambis, Ķīpsala neighbourhood, Riga
1996– (pp. 184–209)

Branch Building of Bank of Latvia
Arhis
Bezdelīgu Str. 3, Kronbergs, Kārkliņš and
Partneri
1997–2001

Z Towers
NRJA (project design), JAHN facade
design, SZK and partners (interior design,
surroundings)
Daugavgrīvas Str. 9/11
2004–

SZK and Partners

Riga Concert Hall

AB Dambis, Riga

SZK合伙人建筑师事务所

里加音乐厅

里加，阿布达比斯

01

A narrow site on AB Dambis a former flood control structure on the left bank of Daugava river just opposite the medieval core of the city, was chosen, and a closed competition was organized in 2006 inviting five Latvian and five well-known European offices: Henning Larsens Tegnestue A/S (Denmark), CoopHimmelb(l)au (Austria), Kada Wittfeld Architektur (Germany), Snøhetta (Norway) and Behnisch, Behnisch & Partner Architekten (Germany). The task was to design a contemporary performance place and a home for three professional organizations – the Latvian National Symphony Orchestra, the State Choir Latvija and the chamber orchestra Rīgas kamermūziķi. Our winning proposal "Lineamentum" envisioned a linear cluster of dark-color logs arranged in a sculptural composition open for expansion. Bearing in mind that the building must perform several varied functions (concerts of different character, office premises for three organizations on weekdays, commercial spaces), it is mandatory that the architectural form be designed by dividing it into different modules expressive of its separate capacities. This not only ensures maximum logic of function and technical solutions, but also creates a sense of scale suitable to the vision of Riga's center and the environs of AB Dambis.

The structure of clearly divided public spaces, three performance halls and staff areas with auxiliary premises was located on the northern part of the dambis connected to the left-bank mainland by a new bridge for cars and a smaller bridge for pedestrians on the very north of the dambis. The diverse orientation of the logs' facets and the finish in reflective, dark tinted glass and ceramic panels ensures color variations of elevations in changing natural lighting, while at the same time maintaining the wholeness of the structure. The use of classical colors on the facade and inside halls guarantees the anticipation of an unforgettable, festive occasion.

The division of the building construction into different modules ensures access to daylight for all spaces in which it is mandatory, its projected situation and dimensions achieving maximum effect through this architectural enhancement. The proposed project development fulfills all demands for optimal function by variants in programming requests, among these that all three concert halls hold stages that are at the same level, with a service platform that can be raised. Functionally, the building is divided into two parallel zones on the pier. On the side toward the panorama of Riga, a shared lobby is planned where streams of visitors can be channeled into hall entrances, cafes, restaurants, conference space, etc. On the Kliversala side, a wing is planned with a gallery on the second floor that joins the administration block and provides a separate entrance for staff and artists in the work spaces.

Credits and Data

Project title: Riga Concert Hall
Client: Ministry of Culture of Republic of Latvia
Location: AB Dambis, Riga
Architect: SZK and Partners
Project team: Andis Sīlis, artist Holgers Ēlers, architects Aleksejs Birjukovs, Ilze Miķelsone, Kristīne Grava, Renāte Pablaka, Indulis Venckovičs, Aigars Kokins.
Acoustic consultant: Andris Zabrauskis
Engineering consultants: Uldis Pelīte and Pauls Sirmais

pp. 164–165: Image of general view from the north. This page: Model for the volume.

第 164-165 页：自北侧看到的全景。本页：建筑体块模型。

阿布达比斯位于道加瓦河左岸，正对里加在中世纪时期的市中心。这里的一处狭长地带曾经是一个防洪构造。2006年，五家拉脱维亚的建筑师事务所以及五家著名的欧洲建筑师事务所，包括亨宁·拉森斯·特格内斯特建筑师事务所（丹麦）、库菲米尔布洛建筑师事务所（奥地利）、卡达·维特菲尔德·阿基尔德建筑师事务所（德国）、斯诺赫特建筑师事务所（挪威）以及贝尼修与合伙人建筑师事务所（德国），围绕此地展开了一场竞赛。竞赛的任务是设计一座当代剧院，它也将成为拉脱维亚国家交响乐团、拉脱维亚国家合唱团以及里加卡梅兹室内管弦乐团三个专业乐团的常驻地。最终的优胜方案“线性方案”提出了一个由深色原木组成的线性集群，并按照具有扩展性的雕塑形式排列。考虑到建筑物必须能够承载不同功能，例如能够举办不同性质的音乐会、方便三个乐团的日常办公以及用作商业空间，建筑形态必然按照不同的用途分别设计。这不仅是为了保证能获得最合理的功能及技术方案，还创造出一种与里加市中心以及阿布达比斯地区环境相适应的尺度感。

有明确结构划分的公共空间、三个剧场以及包含附属设施的工作区位于达比斯的北侧。这里有一座通向市区左岸的机动车桥，最北部还有一座步行桥。朝向不同的原木表面及其反光，加上深色玻璃、陶瓷板，使得建筑立面在自然光下呈现出不同的颜色变化，同时也保持了建筑的整体性。立面和大厅内部使用了古典色彩，为了给观众留下难忘的印象。

通过将建筑划分为不同模块，可以将阳光引入所有有需要的空间，同时也可以利用建筑张力来实现采光及阳光射入量的最佳效果。这一提案满足了项目对于建筑功能的所有需求。三个剧场的舞台处在同一水平面上，同时剧场的服务台可以升降。功能上，整座建筑在栈桥上被分为两个平行区域。方案在面向里加市区的一侧规划有一个共享大厅，客流可以通过这里进入剧场、咖啡馆、餐厅、会议室等空间。在克利弗萨拉一侧规划有一座翼楼，其二楼是画廊。翼楼与行政区域相连，为工作人员及艺术家提供了专用入口。

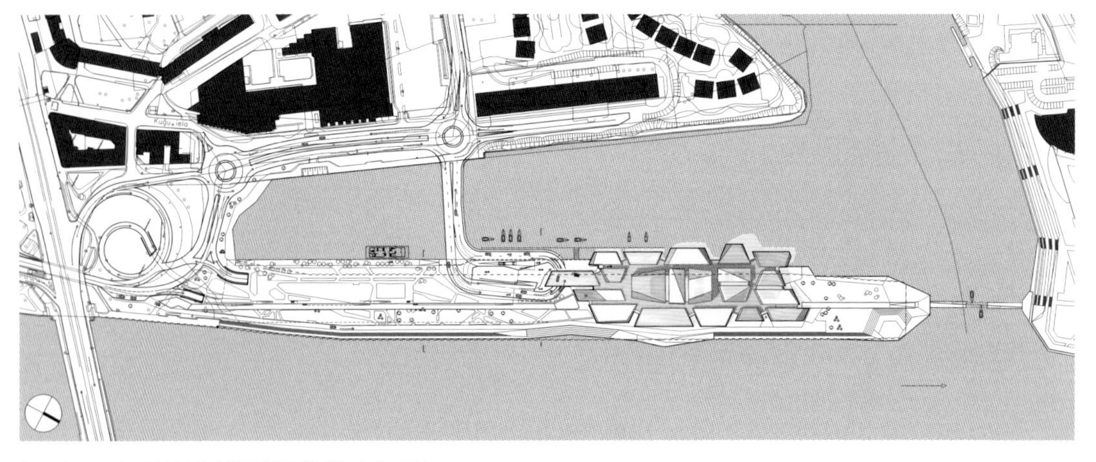

Site plan (scale: 1/6,000)／总平面图（比例：1/6,000）

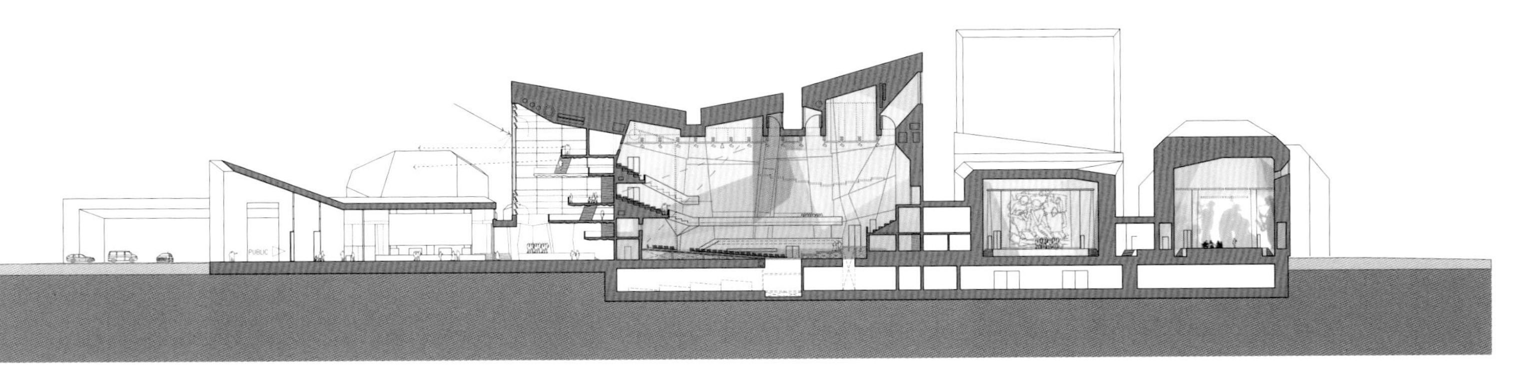

Section (scale: 1/1,200)／剖面图（比例：1/1,200）

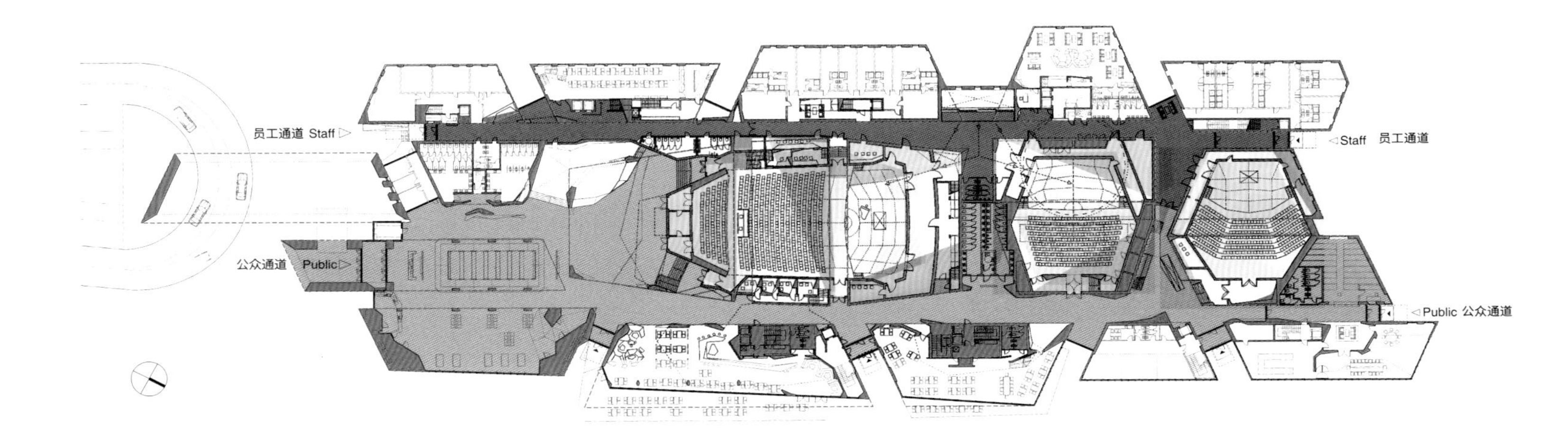

Plan (scale: 1/1,200)／平面图（比例：1/1,200）

Gunārs Birkerts

Latvian National Library

Mūkusalas Str. 3, Zemgales Neighbourhood, Riga 2014

冈斯·伯克茨

拉脱维亚国家图书馆

里加，泽姆盖尔斯社区木库萨拉斯街3号 2014

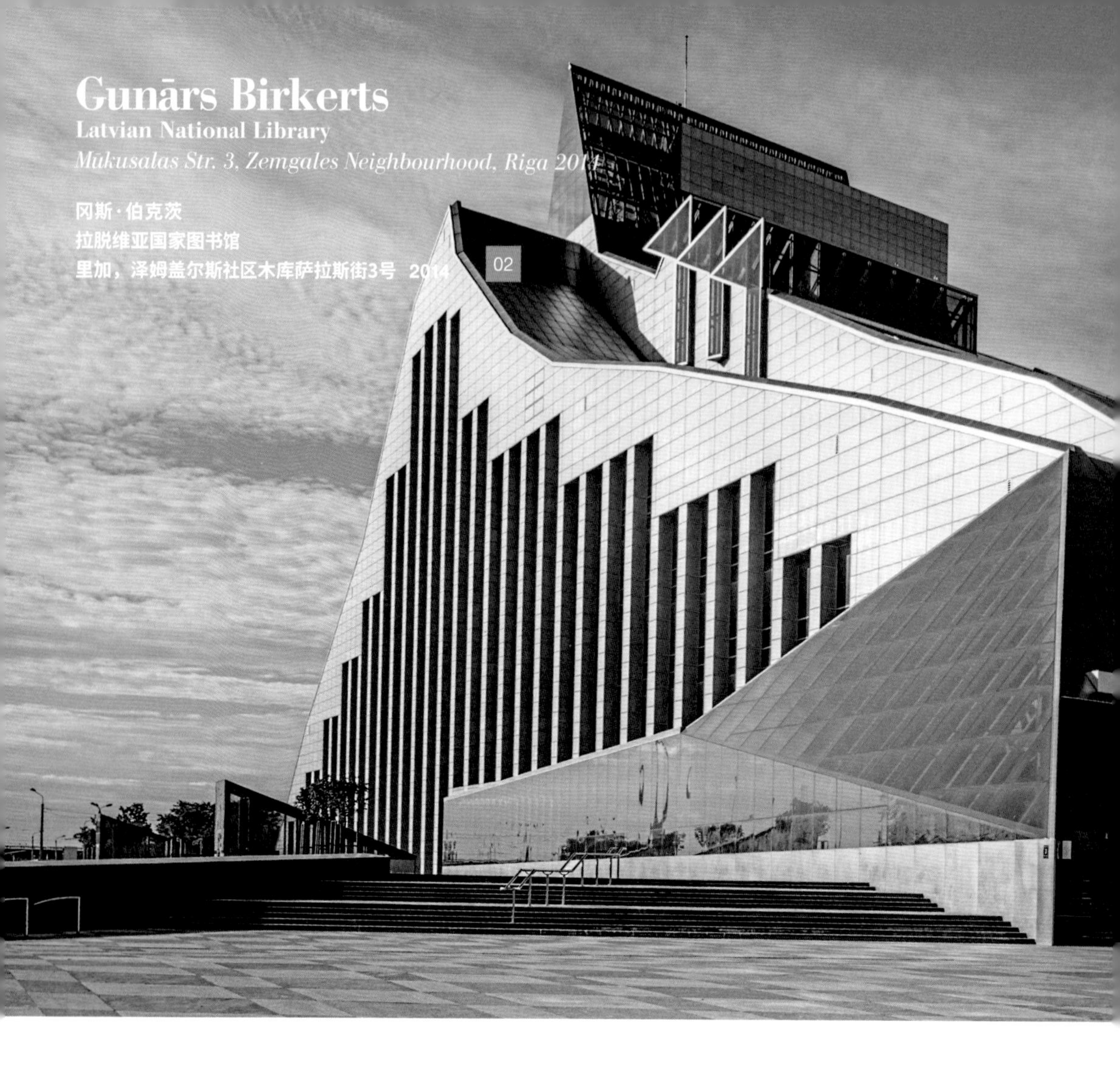

There are lots of symbols and metaphors in my works. The library building includes two of them, the Glass Mountain and the Castle of Light, which are rooted in national culture. It is a dynamic structure with its own face and personality. It is a building that should be understood. The structure is not a visual surprise; it is a harmonious symbiosis of both metaphors. I represent the trend of metaphoric modernists of my generation and some of my works are restrainedly expressive. I studied architecture in Germany and drew my inspiration from the great Scandinavians – Alvar Aalto, Gunnar Asplund, Eero Saarinen. My strength does not come from the big America, no, it comes from Latvia. We have all been involved in the construction process of the Library for 20 years, we have closely followed the evolution of technology and form.

Credits and Data

Project title: Latvian National Library
Client: Latvian Ministry of Culture in collaboration with the Latvian National Library
Construction: 2008–2014
Architect: Gunnār Birkerts in partnership with the office Arhitektu birojs Ģelzis – Šmits – Arhetips
Project team: Modris Ģelzis, Mārcis Mežulis, Dainis Šmits, Sandra Laganovska and others
Structural engineers: LERA, Leslie E. Robertson Associates Structural Engineers
Height: 68.3 m
Length: 170 m
Width: 44 m
Total area: 40,455 m²
Construction costs: 179.7 million euros

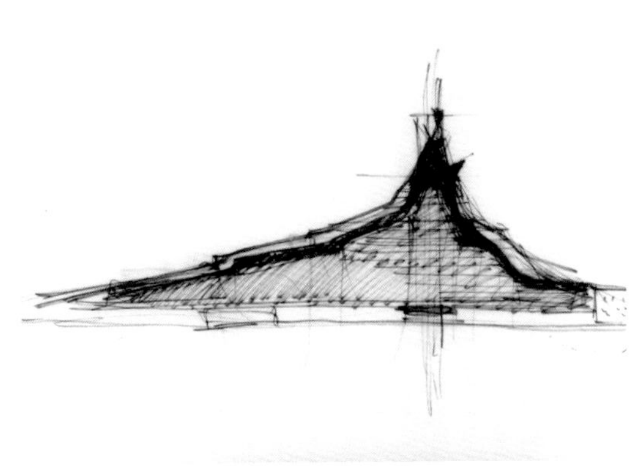

Opposite and this page, above: General view from the north. Photo courtesy of Jānis Lejnieks. This page, below: Sketch of the concept.

对页及本页，上：自北侧看到的全景；本页，下：概念草图。

在我的作品中有很多象征和隐喻。拉脱维亚国家图书馆即由玻璃山和光之城两座建筑组成，它们深深根植于本土的民族文化。它们是动态的结构，有着自己的风貌和个性，需要被人们解读和体会。从视觉上来看，这一结构并不值得惊讶，但事实上，它们代表着两种隐喻的和谐共生。虽然我的部分作品表现力有限，但我也代表着当代隐喻现代主义的潮流。我曾在德国学习建筑，并从阿尔瓦·阿尔托、冈纳·阿斯普隆德、埃罗·沙里宁等伟大的北欧建筑师身上汲取灵感。我的力量并非来自强大的美国，而是来自拉脱维亚。我们都参与了这座图书馆长达20年的建造过程，也密切关注着技术和形式的变革。

Zaiga Gaile Office

Gypsum Factory Renovation

Balasta dambis 70, Ķīpsala neighbourhood, Riga 2004-2015

扎格·盖雷建筑师事务所

石膏工厂改造

里加，普萨拉社区巴拉斯塔达比斯70号 2004-2015

p. 172, above: View of the Phase 2 plaza. The Red House seen at the back is the final volume of the factory project. below: General view from the northeast.Photo by Ansis Starks, courtesy of the architect. p. 173: View of the Phase 1 plaza. The factory chimney in a corner of the plaza was preserved as a monument. p. 174: Interior view of a Phase 2 residence.

第172页，上：二期广场。背景中的红房子是整个项目中最后的工程；第172页，下：从东北方向看到的全景。第173页：一期广场。作为纪念，广场一角的工厂烟囱被保留下来。第174页：二期住宅内景。

Phase I

Historically the gypsum factory was owned by the German merchant Bëhm. The factory is a typical example of the brick architecture of Riga at the end of the 19th century with splendid brick details and textures, interesting wooden constructions, and characteristic window rhythm and division. The location of the factory on the bank of river Daugava can be explained by the convenient access to water for transportation of raw materials and products. A gypsum washing pit had remained in the middle of the courtyard and now the spot is marked with a cast iron water pump. The factory operated until World War II. During the Soviet times the buildings contained laundry facilities for the armed forces.

During the first phase of renovation (2002–2004) one part of the factory was converted into a modern residential ensemble – the first loft complex in Latvia. This part of the factory ensemble consists of five buildings located around a courtyard with a big chimney. The factory offers residential space of different types, from small one bedroom studios to spacious separate houses with three or four levels. Most of the apartments have outdoor space – French balconies, roof terraces, verandas, wooden decks. Apartments are equipped with kitchens and bathrooms, built-in wardrobes and cloakrooms.

Interior design preserves the rough textures and patina of the original brick, metal and wooden constructions. Mainly natural materials have been used, such as old pine plank floors, hexagonal concrete tiles specifically designed for the factory and used in entrance halls, kitchens, terraces and staircases, as well as wooden window frames, windowsills, doors, metal staircases and banisters.

The ensemble contains a boat pier, a restaurant, sauna and underground parking, as well as a floating concrete platform on the bank of the river for the use of the restaurant.

Phase II

Gypsum factory renovation phase II is a logical follow-up to phase I – the aim of the project was to complete the architectural ensemble of the former gypsum factory. The project proposes to renovate the existing valuable historical buildings and to introduce within the scenery of the riverside and within the context of the factory ensemble a significant new architectural volume. The gypsum factory ensemble manifests rehabilitation and humanization of industrial brick architecture. The loft principle is encoded both within the common composition of the ensemble and within the structure of separate residential units. It can be recognized as a search for a new generation versions for living space.

The ensemble consists of four buildings: the Old House, four Villas, the Ogļu Street House and the Red House. The buildings are located around two inner courtyards, one of them is designed as green garden. Colors of the ensemble are pale yellow, red and black. Variations and combinations of the color palette are used throughout the design of the common image. The new buildings are made of bricks which were specifically produced in Switzerland for this project. The same bricks are also used in the interiors of the apartments. All apartments were planned to be sold or rented as design apartments with interior design finish.

Credits and Data

Project title: Gypsum Factory Renovation
Architect: Zaiga Gaile Office
Location: Balasta dambis 70–72, Ķīpsala, Riga, Latvia

PHASE I
Client and developer: Māris Gailis un kolēģi
Project dates: 2000–2004
Construction: 2002–2004
Project team: Zaiga Gaile, Liene Griezīte, Iveta Cibule, Ģirts Kalinkēvičs,
Ingmārs Atavs, Andra Šmite
Collaborators: Gints Lūsis-Grīnbergs, Gints Vaivars
Renovator: Māris Līdaka
Water pump: Sculptor Gļebs Panteļejevs

PHASE II
Client and developer: Domuss Ltd.
Project dates: 2004–2013 / 2013–2015 (Loft)
Construction: 2007–2013 / 2013–2015 (Loft)
Project team: Zaiga Gaile, Liene Griezīte, Iveta Cibule, David Moed, Ingmārs Atavs, Agnese Sirmā, Zane Dzintara, Ineta Solzemniece – Saleniece, Dāvis Gasuls (Loft: Zaiga Gaile, Dāvis Gasuls, Kristīne Riba, Zane Dzintara)
Collaborators: Andra Šmite, Māris Līdaka, Gints Lūsis-Grīnbergs
Total area of the site: 5,022 m² / 670 m² (Loft)
Construction site area: 3,256 m² / 1,192 m²
Number of apartments: 63
Total area of the apartments: 7,807 m²
Basement: 4,300 m²
Underground parking for 80 cars

一期

历史上，这家石膏工厂归德国商人贝姆所有。它是19世纪末里加砖造建筑的典型代表，拥有精致的砖瓦细节和纹理、有趣的木结构以及独具特色的窗户布局。工厂位于道加瓦河河畔，便于利用水路运输原材料及产品。其院子中央的石膏清洗池被保留下来，并用铸铁水泵标记了位置。这家工厂直到第二次世界大战时仍在运营。在苏联时期，工厂中还有供军队使用的洗衣设施。

在一期改造中（2002–2004），工厂的一部分被改建为现代集体住宅，这也是拉脱维亚首个阁楼式复合住宅，由五栋建筑组成，它们围绕着一个带有烟囱的庭院分布。这个改造的工厂能够提供各种不同类型的住宅，既有一居室，也有三至四层的宽敞房屋。大部分的住宅都有法式阳台、屋顶天台、露台、木甲板等室外空间，并具备厨房、浴室、内置衣柜和衣帽间等设施。

室内设计保留了原有砖块、金属及木结构的粗糙纹理和外观。改造主要使用了天然材料，例如旧松木地板，专门为工厂设计的六边形混凝土砖（主要用于门厅、厨房、阳台及楼梯），以及木制窗框、窗台、门框和金属楼梯、栏杆。

这一集合住宅还包含有一座码头、一家餐厅、一家桑拿房以及地下停车场。餐厅还沿着河岸设置了一个混凝土浮台。

二期

二期改造是一期方向的延续，旨在将原石膏工厂改建为集合性建筑。项目计划对现存的珍贵历史建筑进行翻修，同时在滨河景观和工厂建筑群的环境中引入新的建筑体量。石膏厂建筑群体现了砖造工业建筑物的再生及其人性化转变。复式结构原则在建筑群整体构成和独立的住宅单元中均有体现。这可以看作是对新一代居住空间的探索。

整个建筑群包括老住宅、四栋别墅、奥古街住宅和红色住宅。它们围绕着两个庭院而建，其中一个庭院被设计为绿色花园。建筑群以浅黄色、红色和黑色为主色调。为了凸显整体性，这些颜色的使用既有变化又相互配合。这些新建筑是用瑞士专门生产的砖块建造而成，同样的砖块也用于公寓内部。所有的公寓内部都已完成精装，计划出售或租赁。

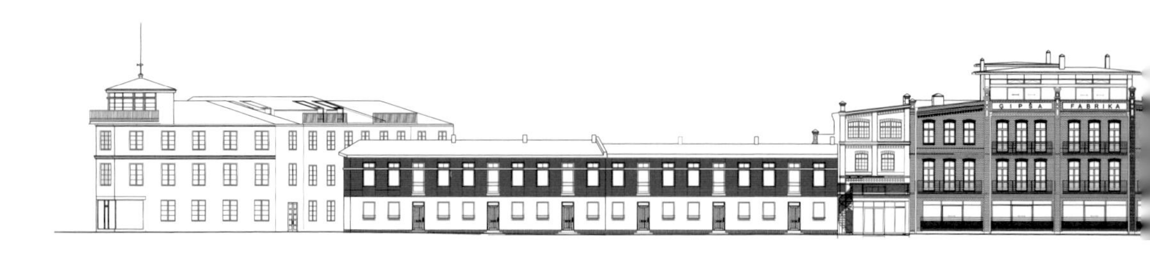

Section (scale: 1/700)／剖面图（比例：1/700）

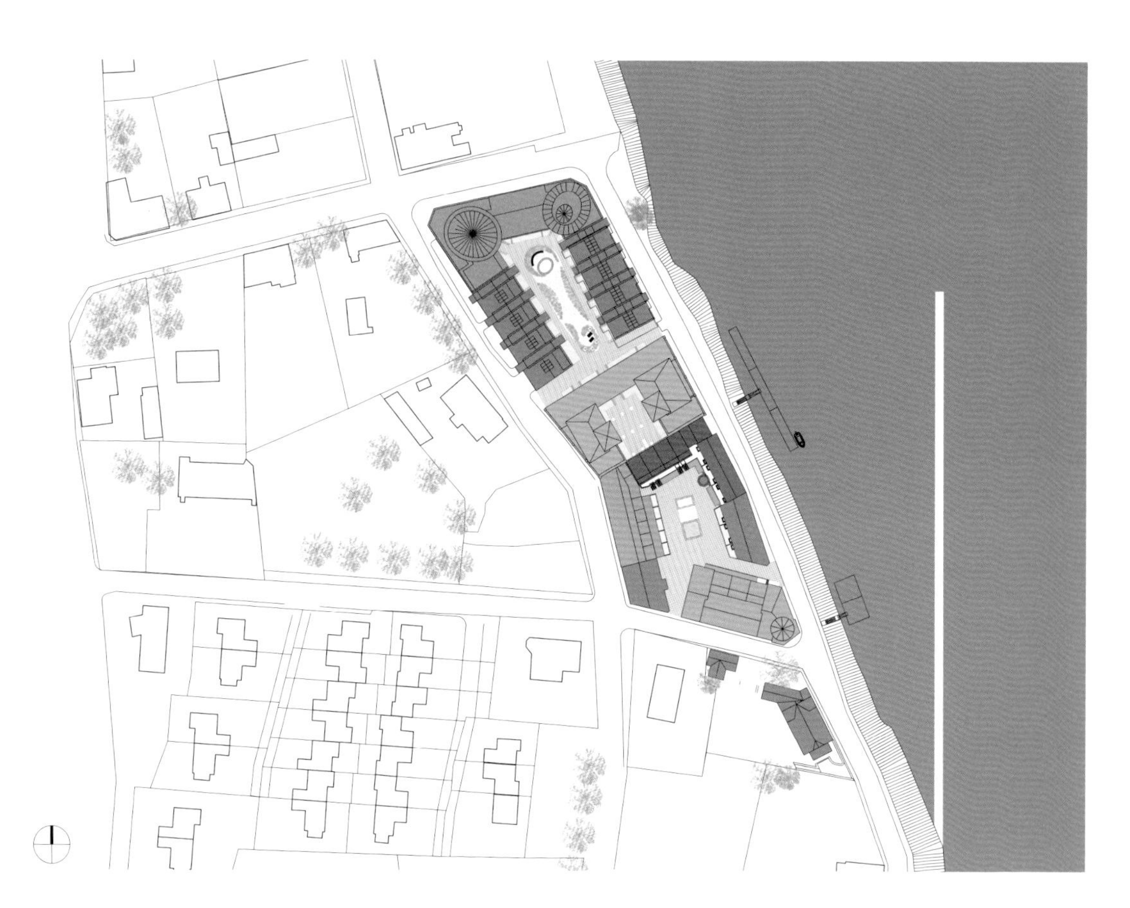

Site plan (scale: 1/2,500)／总平面图（比例：1/2,500）

Zaiga Gaile Office

Žanis Lipke Memorial

Mazā Balasta Str. 8. Ķīpsala neighbourhood, Riga 2012

扎格·盖雷建筑师事务所

扎尼斯·利普克纪念馆

里加，基普萨拉社区 马兹巴拉斯塔大街8号 2012

04

Mazā Balasta street 8 in Ķīpsala, Riga is a place where a Latvian working class couple Žanis and Johanna Lipke saved Jews sentenced to death during World War II by hiding them in a pit under a shed. The Memorial building is the Black Shed – a symbolical shelter under which people were hidden and saved. The image of the building is inspired by the old pitch-black fishermen sheds characteristic of Ķīpsala island. It also spiritually and visually resembles Noah's ark or an inverted boat resting ashore – a symbol of a shelter.

The territory of the museum is surrounded by solid black wooden fence. Passage through the enclosed tunnel that begins by the large entrance gate bears no suggestion of the real scale and structure of the building and it takes a while for the visitor to locate its center with the hidden bunker under the ground.

Visitors are guided along the one-way pathways within the shed, then they ascend to the attic and arrive at the well through which one can see the pit in the basement, designed 3×3×3 meters wide with wooden plank beds resembling the room of the historic bunker. The attic is the main exhibition hall of the museum. Through the roof, sparse sun beams enter the dusky room from the desired freedom outside. Back on the ground floor there is another detour around the bunker. A sukkah is built above the bunker – a symbolic Jewish divine temporary shelter from the cruel world, the desired home with painted transparent paper walls and small windows.

pp. 178–179: Interior of the one-way pathway from the entrance. It consists of planks in the ceilings and upper parts of the walls, concrete with wood pattern imprint on the lower parts of the walls, and a floor of planks and wet polished concrete. Opposite: The building is a wood-frame structure with overlapping wood plank roof and facade. It is passive house and has a heat isolation system that provides a constant low temperature.

第 178-179 页：入口单向通道的内景。通道天花板和墙壁上部使用了木板，墙壁下部使用了木纹混凝土，地板则用木板和湿抛光混凝土建造。对页：这座建筑使用了木构架，屋顶及立面使用了木板。建筑采用了能时常维持低温状态的热隔离系统。

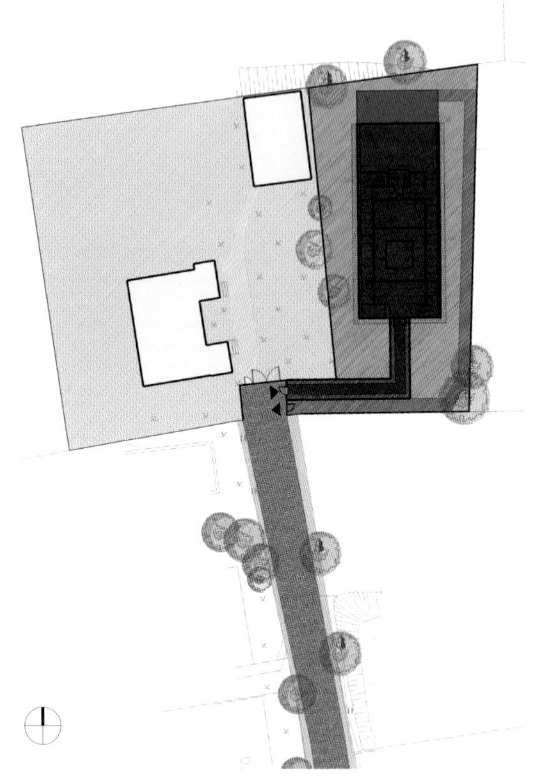

Site plan (scale: 1/800)／总平面图（比例：1/800）

“二战”期间，一对拉脱维亚工人夫妇，让·利普克和乔安娜·利普克，曾将被判死刑的犹太人藏在里加基普萨拉社区马兹巴拉斯塔大街 8 号的一个地窖中。与“黑人避难所”类似，这座纪念馆也是人们被隐匿而得救的避难所的象征。这座建筑的设计灵感来源于基普萨拉岛古老的漆黑渔棚。同时，在精神和视觉上，它也能让人想起诺亚方舟或者一艘海岸上的倒置小船，更加契合避难所的形象。

纪念馆被坚固的黑色木栅所包围。从入口大门开始，游客难以看出建筑物的实际大小和构造。他们首先要通过一条封闭的隧道，然后花费一段时间到达建筑中心。中心的地下就是曾经藏匿犹太人的地窖。

游客被引导着沿室内单行道游览。在登上顶楼后，他们可以看到一口井，并可以透过它看到地窖。这间地窖为 3 米 ×3 米 ×3 米，室内还有木板床，与历史中的形态相似。顶楼是纪念馆的主要展厅。透过屋顶，稀疏的阳光可以进入昏暗的房间，与犹太人渴望自由的愿景呼应。一层的后方，有另一条通向地窖的迂回道路，而地窖上也建造有圣所，采用了绘制的透明纸墙和小窗户，象征着犹太人暂时离开残酷的世界，回到一个温暖的家。

Credits and Data

Project title: Žanis Lipke Memorial
Location: Mazais Balasta dambis 8, Riga, Latvia
Project dates: 2005–2009
Construction: 2008–2012
Architect: Zaiga Gaile Office
Project team: Zaiga Gaile, Ingmars Atavs, Agnese Sirmā, Ineta Solzemniece, Zane Dzintara, Maija Putniņa-Gaile, Dāvis Gasuls
Concept authors: Māris Gailis, Augusts Sukuts, Viktors Jansons
Artist: Kristaps Gelzis, Reinis Suhanovs
Sound: Jēkabs Nīmanis
Project manager: Māris Gailis
Useful area: 395 m²

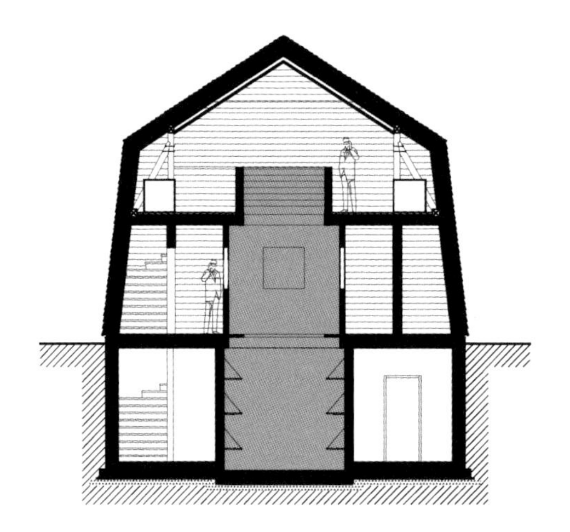

Section (scale: 1/200)／剖面图（比例：1/200）

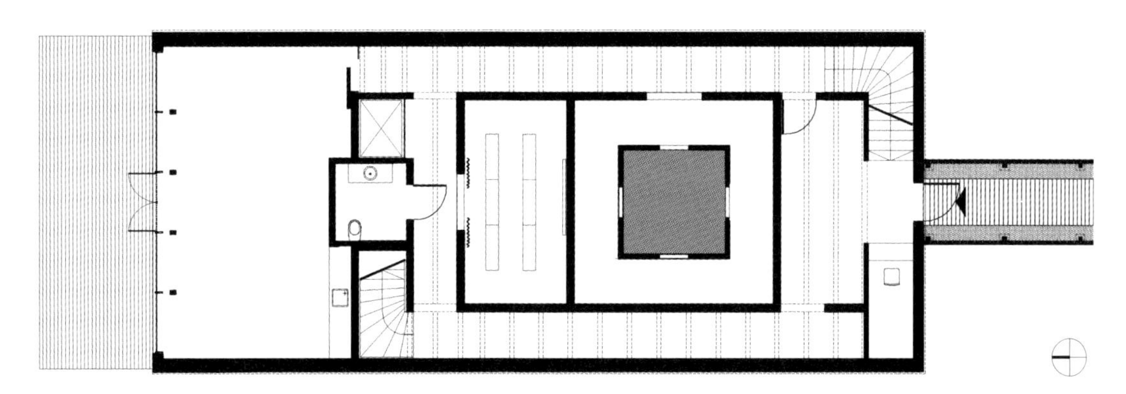

Ground floor plan (scale: 1/200)／一层平面图（比例：1/200）

Zaiga Gaile Office

The Collection of Wooden Houses in Ķīpsala

Balasta dambis, Ķīpsala neighbourhood, Riga 1996-

扎格·盖雷建筑师事务所

基普萨拉岛上的木构建筑群

里加，基普萨拉社区巴拉斯达比斯 1996-

05

Site plan (scale: 1/4,000)／总平面图（比例：1/4,000）

Ķīpsala is a small island in the Daugava river, close to the center of Riga, with a unique character – it has valuable architectural heritage and a remote countryside feeling. The site has long been neglected and therefore has retained many of its historical values. Historically the island has always been inhabited by fishermen. Its original architectural heritage is made up of fishermen's dwelling houses, sheds, fish smoking houses and simple working class apartment buildings. Since 1998 its urban structure has been protected as a monument of state significance.

Over the last 20 years Zaiga Gaile Office has renovated 14 historical wooden houses on the island. With her determined work architect Zaiga Gaile has been able to change the public attitude towards the heritage of wooden architecture in the urban environment, as a result of which the wooden architecture has now been recognized as one of the symbols of the urban structure of Riga.

The latest project, The Collector's House, was finished in October, 2015. It unites architecture of three centuries – wooden house in the Neoclassical style from the 19th century, brick house in the style of Functionalism from the 20th century and the new addition which is the 21st century. It is a sophisticated dialogue between the historical heritage and contemporary lifestyle within the spectacular scenery of the riverside Ķīpsala.

Credits and Data

Project title: The Collection of Wooden Houses in Ķīpsala
Project dates: 1996–
Location: Ķīpsala, Riga, Latvia
Architect: Zaiga Gaile Office

pp. 184–185: Aerial view of Ķīpsala island from the southeast. Wooden houses stand along the riverside. Courtesy of The Museum of the History of Riga and Navigation.

第184-185页：自东南方向鸟瞰基普萨拉岛。木构建筑沿河岸排布。

基普萨拉是位于道尔加瓦河中的一座小岛。它靠近里加市中心，拥有极具历史价值的建筑遗产和远离城市的乡村氛围。这座小岛长期被忽视，但也因此保留了历史风情。历史上，这座岛一直是渔民生活的地方。岛上存留的建筑遗产主要是渔民的住宅、仓库、熏鱼小屋以及工人简朴的集体住宅。1998 年起，这里的城市结构被视作国家历史遗迹得以保护。

在过去的 20 多年里，扎格 · 盖雷建筑师事务所已对岛上的 14 座历史木屋进行了改造。建筑师扎格 · 盖雷凭借自身坚定的意志，改变了公众对于城市空间中木构建筑遗产的态度。现在木构建筑已经被看作是里加城市结构的象征。

事务所的最新项目“收藏者之家”于 2015 年 10 月完工。它融合了 19 世纪新古典主义木构住宅、20 世纪功能主义砖造住宅以及 21 世纪新建筑要素的特征，反映出历史遗产和当代生活方式之间的巧妙对话。

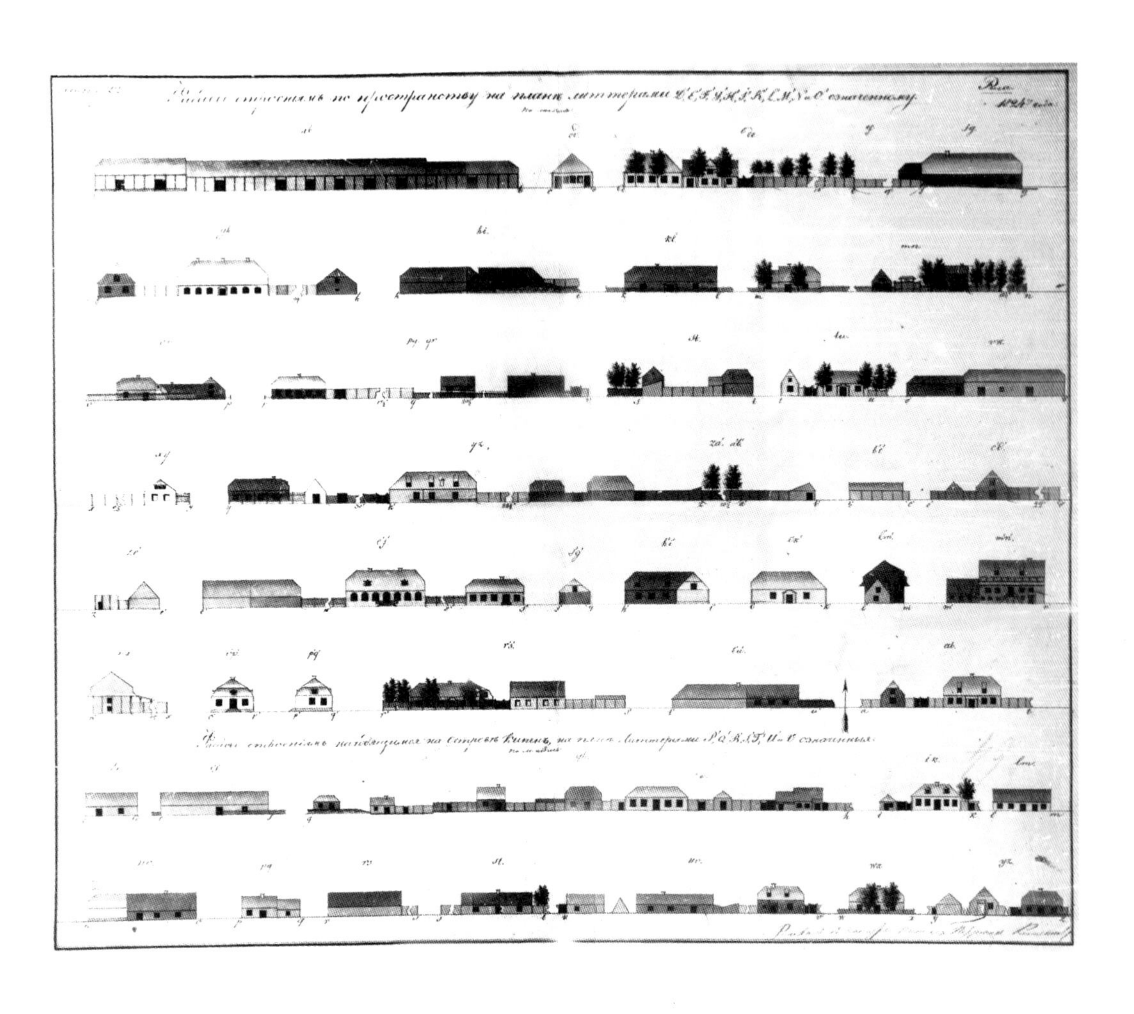

Opposite: Postcard of Ķīpsala from the begining of the 20th century (part). This page: The collection of fasade in Ķīpsala and neighborhood during 1823–1824. These were documented in the order as they appear on the street. Courtesy of Zaiga Gaile Office.

对页：20 世纪初的基普萨拉岛明信片（部分）。本页：1823-1824 年间基普萨拉岛上木构建筑的立面集合。这些都是按照它们竣工的顺序记录下来的。

Baznicas Str. House
2005

巴兹尼卡斯街住宅
2005

This page: General view from the north. The house was relocated from the city center. It is an outstanding monument of the Classical architecture of Riga's mid-19th century city center. Photo by Ainars Meiers, courtesy of the architect. Opposite, above: Exterior view of the facade. This monument of wooden architecture in the Art Nouveau style was built by architect Eižens Laube in 1908 as a house for workers of the local factory. The renovation project converts the small worker flats into spacious modern apartments. Photo by Ainars Gaidis, courtesy of the architect. Opposite, below left: View of the kitchen. Opposite, below right: Close up view of the stairs. Two photos below by Ainars Meiers, courtesy of the architect.

本页：自北面看到的全景。这座住宅是从城市中心移建的，具有 19 世纪中叶里加市中心古典风格建筑的典型风格。对页，上：正面外观。这座具有纪念意义的新艺术风格木构建筑是建筑师艾森·劳贝于 1908 年为当地工厂工人建造的住所。翻修工程将工人公寓改造成宽敞的现代住宅；对页，左下：厨房。右下：楼梯。

Laube's House
2005

劳贝之家
2005

Briana Str. House
2007

布里亚纳街住宅
2007

Anrijs' House
2000

安立吉斯之家
2000

Opposite, above: General view from the north. The history of Baroque style log buildings dates back to the 18th century. The building was relocated from the city center to Kipsala and in the course of renovation regained its original Baroque manor house image and the original layout structure. Opposite, below left: View of the extension's pool. A new extension with veranda, pool and garage in contemporary style has been added to the building. Opposite, below right: View of stairs preserved in original house. Three photos by Ainars Meiers, courtesy of the architect. This page : General view from the west. This log house was built during the second half of the 18th century and is a local architectural monument. Photo by Ainars Meiers, courtesy of the architect.

对页，上：自北面看到的全景。这座巴洛克式原木建筑的历史可以追溯到 18 世纪。该建筑从里加市中心搬迁至基普萨拉岛，并在改造过程中恢复了其原有的巴洛克式房屋的样式及结构；对页，左下：扩建的游泳池。这里增建了一个带有现代风格的阳台、游泳池和车库；对页，右下：原住宅中保留下来的楼梯。本页：自西面看到的全景。这座木屋建于 18 世纪后半叶，是当地一座具有纪念意义的建筑。

Australian's House
2005

澳大利亚人之家
2005

This page: General view from the southwest. The 100-year-old former workers house was converted into a luxury apartment building with two apartments. Before the renovation the house was bought by an Australian who had plans to tear it down and build a new house. The bronze kangaroo atop the building is a reminder of that perilous episode in the history of the house. Photo by Ainars Gaidis, courtesy of the architect. Opposite, above: The extension of the gymnasium is seen at the front next to the extension of the classrooms, with the original building at the back. In 2001 a two-level house built in the 1930s was adapted to the needs of the International pre-school. The school has grown and soon needed a new extension. The new annex is defined as an addition of functional wooden architecture which nevertheless clearly repeats the silhouette of the original house. Opposite, below left: Interior of the gymnasium. Opposite, below right: View of the corridor. Three photos by Ainars Meiers, courtesy of the architect.

本页：自西南看到的全景。这座有着100年历史的旧工人住宅被改建成一座豪华的双户公寓楼。在改造前，这座建筑曾被卖给一位澳大利亚人，他计划将建筑拆除重建。屋顶上的青铜袋鼠塑像能够让人们联想到这段历史。对页，上：前方是扩建的体育馆，旁边为扩建的教学楼，它们后面是原有建筑。2001年，为适应国际学前教育的需要，一座建于20世纪30年代的两层建筑被改造。随着学校的发展，原有建筑需要被扩建。增建部分是功能性木结构建筑的补充，同时也仿照了原有建筑的外观；对页，左下：体育馆内景；对页，右下：走廊。

International School
2003

国际学校
2003

Gatis' House
2005

盖提斯之家
2005

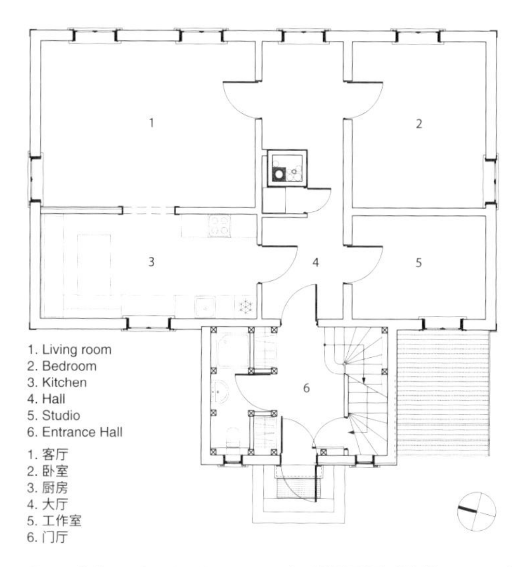

Ground floor plan (scale: 1/200)／一层平面图（比例：1/200）

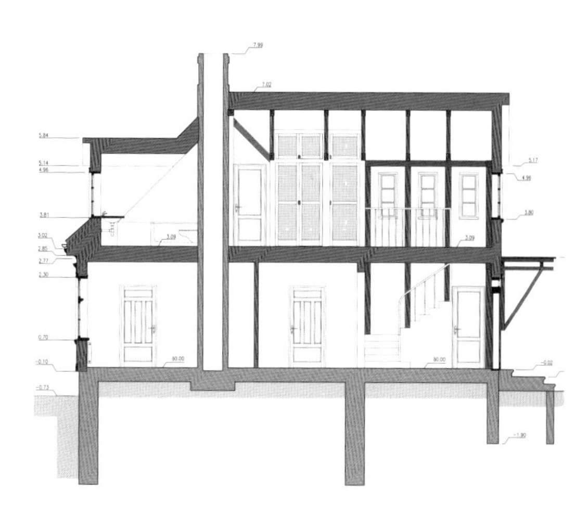

Section (scale: 1/200)／剖面图（比例：1/200）

Opposite: View of the west facade. The site of the building appears in a map of 1790. It is one of the oldest log buildings in Ķīpsala. The second floor which was once an unused attic is a bedroom now. Photos are by Ainars Meiers, courtesy of the architect. This page, 2 photos: General view of the house at present and before. Two photos by Ainars Gaidis, courtesy of the architect.

对页：建筑西立面。1790年，这座建筑就已经出现在了地图上。它是基普萨拉岛上最古老的木构建筑之一。二楼的一个闲置阁楼现在变成了卧室。本页，两张照片：改造前后建筑全景。

Philosopher's House
2003

哲人之家
2003

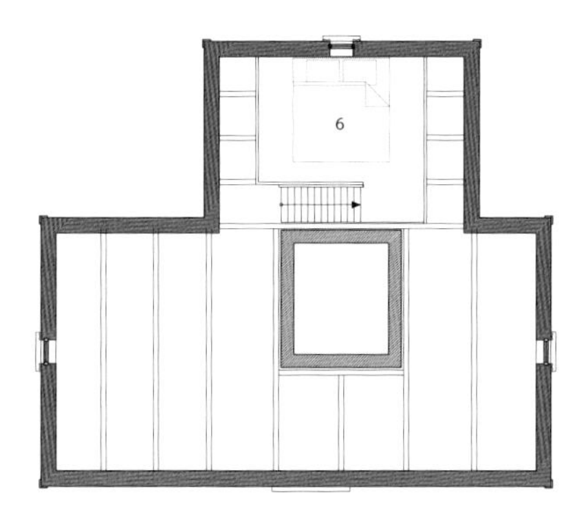

Second floor plan／二层平面图

Ground floor plan (scale: 1/200)／一层平面图（比例：1/200）

Opposite: Close up view of the facade. The house is traditional wooden architecture typical of Ķīpsala and has been renovated using the traditional construction methods and materials: for the facade wood tar, flour based natural paint and linseed oil paint, and for the roof tar paper. Photos by Ainars Meiers, courtesy of the architect. This page, 2 photos : General views of the house at present and before. Two photos by Ainars Gaidis, courtesy of the architect.

对页：建筑立面。这座房子是基普萨拉岛典型的传统木构建筑。在改造过程中，建筑师也采用了传统的建造方法和材料：外墙使用了木焦油、小麦基自然涂料和亚麻油涂料，屋顶使用了焦油纸。本页，两张照片：改造之后和之前的房屋。

This page: View of the living room. A chimney is seen in the center. Lime plaster and adhesive paint are used for the interior finish. Photos by Ainars Meiers, courtesy of the architect.

本页：客厅。客厅中央有一座烟囱。内部装修使用了石灰膏和胶漆。

Collector's House
2015

收藏者之家
2015

This page: General view from the east. The house's name refers to a collection of three architectural styles. Photo by Ansis Starks, courtesy of the architect.

本页：自东面看到的全景。由于这座建筑是三种不同建筑风格的集合，所以被命名为“收藏者之家”。

Gailis' House
1997

盖里斯之家
1997

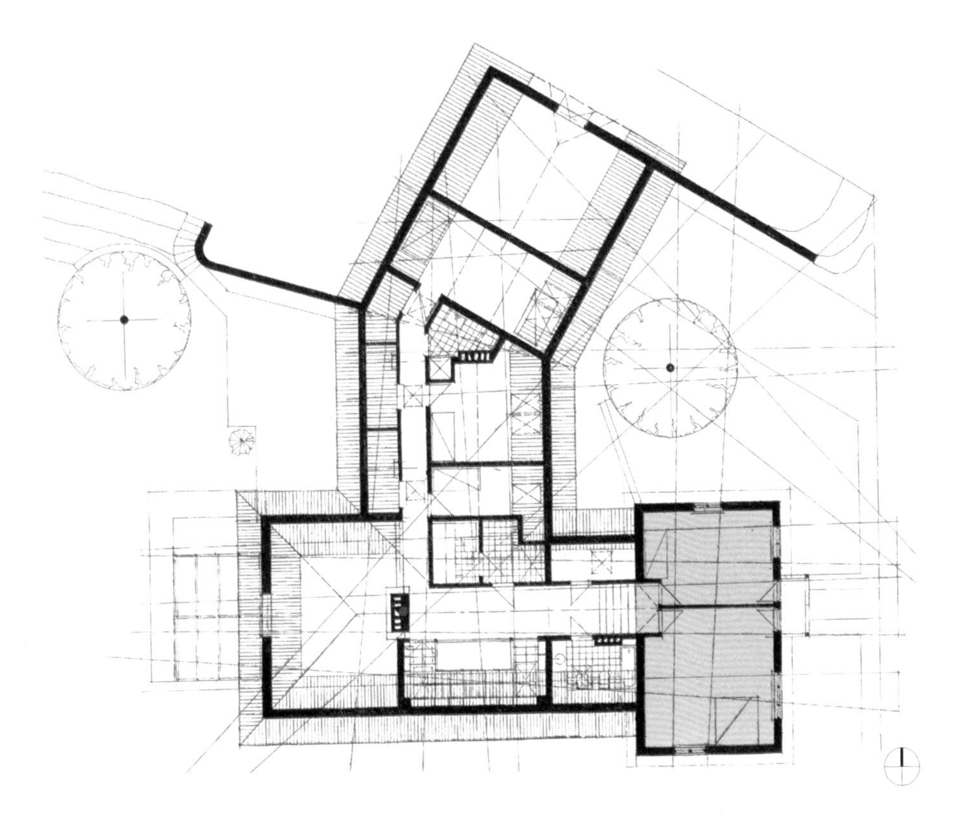

Ground floor plan (scale: 1/300)／一层平面图（比例：1/300）

Opposite: View of the west facade.

对页：建筑西立面。

pp. 206–207: Interior view of the living room. The light of the garden come into the room. On the left, stair with book shelves and skylight above it are seen. This page, above: View of the entrance. This page, below: Interior of the bed room on the second floor.Opposite: View from the garden. The spacious garden with various native flowers was once a fisherman's vegetable garden.

第 206-207 页：起居室内景。阳光从花园照进房间。左边可以看到楼梯、书架和天窗。本页，上：入口视角；本页，下：二楼卧室内景。对页：花园。宽敞的园子里种植有各类原生植物，这里曾经是一位渔夫的菜园。

Interview 1:

Architectural context in Latvia and Riga

Interviewer: Ilze Paklone

访谈 1

拉脱维亚与里加的建筑脉络

采访者：伊尔泽·帕克罗内

In your opinion, what are important architecture or historical landmarks in Latvia or, more precisely, in Riga?

Our source code is the traditional farmsteads and dwelling houses of farmers. In that source code one can read the solitary nature of their lifestyle. The farmers in Latvia never really lived in villages, but rather in farmsteads that are complete building complexes themselves for a family, those complexes in turn creating clusters. Villages are more evident in Latgale, the Eastern part of Latvia, where the Slavic tradition has had more influence, or along the seashore as fishing requires joining workforces.

The characteristic aspect of the solitary farmstead complex is that for every need of the farm a separate house would be built. The center of the complex is an oak or lime tree. This attitude encodes the ensemble mode of thinking. That is a very clear way of living.

The archetypical form of the house is the pyramidal, with monumental roof with large roof overhangs. The shape has derived entirely from our landscape and climate as we have to consider multiple aspects of climate – cold and hot, humidity, snow – and to protect the wooden walls from them.

Buildings in the Ķīpsala ensemble in Riga are transitions in lifestyle between countryside and the city. There are houses with vast gardens – ideal places for living at the river. In fact, all Pārdaugava (part of the city on the left bank of the river Daugava) has this rural-urban structure of straight streets as in the city and with silent gardens as courtyards behind the houses.

Zaiga Gaile, Zaiga Gaile Office

It seems to me that the city of Riga is special because it sparks a feeling of being part of the world's events and having dense layers of history despite its small scale. Riga has a provincial

在您看来，拉脱维亚，或者更确切地说，里加有哪些重要的建筑或具有历史意义的地标？

我们的原点是传统的农场和农舍，你可以从中读出他们生活方式本质中的独立。拉脱维亚的农民从未真正住在村庄之中，对于一个家庭而言，农场就是完整的住宅，多个农场则会形成聚落。村庄在拉脱维亚东部的拉特加尔更为明显，那里更多受到斯拉夫传统的影响，同时沿海地区的渔业也需要更加密集的劳动力。

独立农场的特点在于，每个需求都会带来一栋相应的房屋，建筑群的中心往往是一棵橡树或椴树。它基于一种从整体出发的思维模式，生活形态一目了然。

农舍建筑主要采用金字塔形，屋顶厚重，挑檐宽阔。之所以呈现这种形态，是基于这里的地形和气候，我们不得不考虑寒冷、酷热、潮湿、降雪等各种气候条件，以保护木质墙体不受威胁。

里加基普萨拉岛上的建筑群则体现了一种介于农村和城市之间的生活方式。那里的房屋都附带大花园，可谓沿河居住的理想之所。事实上，整个帕道加瓦（里加在道加瓦河左岸的城区）都是这种城乡结合式的布局，既有城市常见的笔直街道，也有藏在住宅背后的安静庭院。

——扎格·盖雷，扎格·盖雷建筑师事务所

对我来说，里加这座城市之所以特别，是因为它能激发一种归属世界的情感，尽管面积不大，却有着深厚的历史积淀。里加兼具本土性和国际性，换言之，这是一座拥有大都会特质的小城。这或许很难理解，但原因在于，这个特殊的地理空间被众多文化包围，我们仿佛始终处在人来人往的十字路口。当不同文化背景的人们走过，便给里加留下了当代大都会的气息。

——安德里斯·克朗博格和雷蒙兹·萨乌提斯，
阿尔希斯建筑师事务所

and international appearance at the same time. This is something almost incomprehensible. A small city with a metropolitan character. The reason for this is the specific geospatial status of being enfolded in-between many cultures that eventually implanted the feeling that we are always at the crossroads of flows of people coming and going with manifold aspects of culture. As a surplus – it is a contemporary metropolitan feeling.

Andris Kronbergs and Raimonds Saulītis, Arhis

Rather than mentioning a single building, I would say that outstanding urban structure is the 19th-century semicircle of boulevards with green parks on both sides of the City Canal. It is the result of an ambitious large-scale urban project that involved tearing down the fortification wall system and envisioning an entirely new expansion of the city. It is something to enjoy every day in the city. That is powerful. And, it is also part of the common European cultural heritage. Construction logic can be observed and traced in the city. The buildings have been there for such a long time, it is worth contemplating over them as some kind of force. Of course, we also somehow always relate to the idea of farmsteads in the countryside.

In fact, historical opportunity and our ability to spring out of the farmstead lifestyle into the relatively aristocratic urban lifestyle with its prototypical 19th century apartment house with its typical 21.3-m cornice, high ceilings and 24-m high ridge of the roof can also be considered an encouraging historical moment. Also the idea of urban blocks clustering the apartment houses encodes a prototypical character of the city.

Perhaps, this transition also spotlighted the desire to enjoy urban lifestyle. This definitely is a source of inspiration.

In essence, we have the opportunity in Riga to enjoy both Eastern and Western cultures. That is remarkable.

Reinis Liepiņš, Sudraba Arhitektūra

要说这座城市出色的结构体，与其提名某一栋单独的建筑，我更倾向于19世纪建成的、城市运河两侧穿过公园的半圆形林荫道。它始于一个雄心壮志的大型城市项目，需要拆除成片的护城墙，还设想了一个全新的城市扩张方案。它能装点城市居民的日常生活，而且是欧洲共同文化遗产的一部分。观察城市、追溯历史，不难发现它建造的逻辑。城市中的一些建筑存在了很久，足以被视为某种力量。当然，我们总能找到它们与乡村农场之间的关联。

事实上，无论是历史机遇，还是我们自身从农场到相对贵族化的生活方式的跃迁（19世纪的公寓楼就是典型，有着21.3米长的檐口、高高的天花板、24米高的屋脊），都堪称推进历史的时刻。将公寓楼聚集为街区的想法也体现了城市的特征。或许，这种转变还凸显了人们对享受都市生活的渴望。但不管怎样，它是灵感之源这一点毋庸置疑。

归根结底，在里加，我们有机会同时欣赏东西方文化。这是具有标志性的特点。

——雷尼斯·列平修，苏德拉巴建筑师事务所

Non-urban Environments

非城市地区

An aspect that shapes an important part of what could potentially be termed as contemporary Latvian-ness is a lifestyle that always fluctuates between urban sophistications and a non-urban, rustic approach. Contemporary second homes in the rural areas of Latvia or just at the fringes of Riga are contemplative hideaways from the urban rush, being also a distant echo of the traditional solitary farmstead ensembles surrounded by trees. Cultural venues from other, smaller cities in Latvia also feature important and conscious attempts to show that Riga is not the sole center of cultural sophistication.

所谓的当代拉脱维亚风格，或许是在城市洗练的生活方式与非城市朴素风格的不断波动中被塑造的。位于拉脱维亚农村或里加周边的现代别墅，是让人远离都市喧嚣、心情平静的隐居之所。这里被树木包围，与传统孤立的农场建筑群相似。同时，拉脱维亚其他小城市中的文化场所也具有重要价值，这说明里加并非这个国家唯一的文化中心。

Architecture in the Non-urban Environments of Latvia／拉脱维亚非城市地区建筑分布

8 BLACKS
NRJA
Rumba parish, Kuldīga
2011 (pp. 218–223)

2 SISTERS
NRJA
Langstiņi village, Garkalne
2009 (pp. 224–227)

House of Ruins
NRJA
Saka parish, Pāvilosta
2006 (pp. 228–231)

Vacation Home on Easter Island
Zaiga Gaile Office
Roja parish, Roja
2010 (pp. 232–237)

Saldus Music and Art School
MADE arhitekti
Saldus town, Saldus
2013 (pp. 238–243)

View Terrace and Pavilion
Didzis Jaunzems, Jaunromāns un Ābele
Krievkalna island, Koknese parish, Koknese
2013 (pp. 244–249)

Latvian National Open Air Stage
Mailītis A.I.I.M.+ Architect J. Poga Office
Mežaparks neighbourhood, Rīga
2007–2022 (pp. 250–255)

Talsi Creative Yard
MARK arhitekti
Talsi town, Talsi
2013

Pārventa Library
INDIA
Tārgales 4, Ventspils city
2009

Pavilion and Workshops for Nature Concert Hall
Didzis Jaunzems
Sigulda town, Sigulda
2014

Dzintari Concert Hall Reconstruction
Jaunromāns un Ābele, Ināra
Heinrihsone
Turaidas Str. 1, Jūrmala city
2016

Concert Hall Liepaja "Great Amber"
Volker Giencke & Company
Radio Str. 8, Liepāja city
2015

Centre of Creative Services of Eastern Latvia "Zeimuls"
SAALS
Krasta Str. 17, Rēzekne city
2014

Majori Primary School Sports Hall
Substance
Rīgas Str. 1, Jūrmala city
2008

Riga International Airport extension
Arhis
Mārupe
1998–2001

Printing House "Britania"
Arhis
Daugmale parish, Ķekava
2007

Single Family House
Arhis
Zemeņu Str., Jūrmala city
2007

Holiday Home "Lapiņas"
Brigita Bula Architekt
Roja parish, Roja
2011

Rūmene Manor renovation
Zaiga Gaile Office
Kandava parish, Kandava municipality
2004–

Cēsis Brewery reconstruction for Art & Science Center
Mailītis A.I.I.M.
Lenču Str. 9/11, Cēsis town, Cēsis municipality
2015–2025

Photo Credit:
02: Photo by Gatis Rozenfelds, courtesy of the architect / 03: Courtesy of the architect /04: Photo by Ainars Meiers, courtesy of the architect /05: Photo by Ansis Starks, courtesy of the architect /06:Photo by Ansis Starks, courtesy of Maris Lapins /07: Courtesy of Mailtis A.I.I.M. /08 Photo by Gvido Kajons, courtesy of the architect / 09: Photo by Ansis Starks, courtesy of the architect /10:Photo by Ernests Sveisbergs, courtesy of the architect /11: Photo by Ansis Starks, courtesy of the architect /12: Photo by Indrikis Sturmanis, courtesy of Volker Giencke & Company /13: Photo by Janis Mickevics, courtesy of the architect /14: Photo by Martins Kudrjavcevs, courtesy of the architect /15: Photo by E.Matvejeva, courtesy of the architect / 16: Photo by I.Sturmanis, courtesy of the architect/17:Photo by Kristaps Šulcs, courtesy of the architect /18:Photo by Maris Lapins, courtesy of the architect /19: Photo by Ansis Starks, courtesy of the architect / 20:Courtesy of the architect

NRJA
8 BLACKS
Rumba parish, Kuldīga 2011

NRJA
“8座黑色建筑”组群
库尔迪加伦巴区 2011

01

Credits and Data

Project title: 8 BLACKS
Client: Private person
Program: Traditional living complex in Kurzeme
Location: Līči, Kuldīga region, Latvia
Project design time: 2004–2010
Construction time: 2005–2011
Architect: Architectural office NRJA
Project team: Uldis Lukševics, Ieva Lace, Linda Leitane, Ints Mengelis
Building construction: BICP / Gatis Vilks, Viktors Mitrofanovs
Built-in furniture: GSR group
Construction company: RBS SKALS
Site area: 6.58 ha
Built area: 1,358.7 m²
Built volume: 4,681.5 m³
Living spaces: 461.6 m²
Technical spaces: 456.9 m²
Terraces: 215.7 m²

1. Territory gate
2. Access road
3. Guest house
4. Volleyball field
5. Football field
6. Front square
7. Main entrance and outhouse
8. Main house
9. Tower–studio
10. Courtyard
11. Garage
12. Storage
13. Greenhouse
14. Cellar
15. Woodshed
16. Shed
17. Sauna house

1. 庄园大门
2. 入口道路
3. 客房
4. 排球场
5. 足球场
6. 前广场
7. 主入口和室外
8. 主屋
9. 塔楼 - 工作室
10. 中庭
11. 车库
12. 储藏室
13. 温室
14. 地窖
15. 柴房
16. 棚
17. 桑拿房

Site plan (scale: 1/2,500)／总平面图（比例：1/2,500）

Initially planned as a holiday house, at the end of the design process it became the socially active family's living house complex at the River Venta near Kuldīga.

The place drags one in to live there.

8 BLACKS consists of:

1. A group of buildings for daily living, located around a paved courtyard that creates a center of gravity for the complex,

2. A guest house and sauna house located at a distance,

3. A cellar sitting within the topography of the site and ancillary buildings,

4. Football and volleyball fields.

8 BLACKS is a modern approach for a traditional living complex in Kurzeme adopting the archetypal cross-section as a common element for all buildings. The union of traditional and modern in the image of the house entails a sense of inheritance and belonging while at the same time providing the necessary comfort for active living.

The expressive nature of the terrain – a swift river, remarkable topography, wind – defined the minimalist form of buildings; they do not compete with nature. By using one material for both walls and roofs, a homogenous look of the buildings is achieved; surface texture and details are produced by the finish of raw wood planking.

In contrast to 8 BLACKS, the inner courtyard is connected to a garage building made of white polycarbonate that operates as a large lamp, illuminating the yard.

8 BLACKS – 8 black painted wood planking buildings forming a complex.

The large windows of the buildings provide a visual link with the surroundings; the terraces of various sizes offer an opportunity to go out while being in the house. Indoor space is formed following a principle of 'box in a box' – free-standing room blocks with built-in furniture walls. The space above the ceilings is used as extra sleeping area.

A two-story volume separated by a gallery and a terrace from the rest of the residential building contains a fireplace room on the ground floor and the owner's working room on the upper level. Construction of walls in this building is different from other volumes – metal frame construction with glass fill and rare wooden siding creates light games in the interior. Wooden boards, opened up towards the river, allow watching it.

Placed within the existing topography, the cellar volume ends up the planed garden with a solid rock wall. Moreover, the garden within the gully will be protected from prevailing winds. The cellar is located at a distance from the living complex, while a visual link is provided through an opening in the garage building. The final volume of the complex – the gate house – will visually complete the private courtyard; it will place a wide, comfortable main entrance with a place for the dog and a technical equipment room.

The 8 BLACKS identity sign that appears on the main entrance wall is made by modifying the site plan and thereby obtaining the sign of the complex. By multiplication and combination, one gets a drawing, which will be set up on the white wall of the master bathroom.

pp. 218–219: View of the northeast part of site. The sauna is seen in front and the woodshed at the back. Following the tradition of the farmsteads in Latvia, each building has its different function and the total collection makes one household. Photo courtesy of the architect.

第218-219页：场地东北。前面是桑拿房，后方是柴房。仿照拉脱维亚的传统农庄，每座建筑都具备不同的功能，它们共同构成一个完整的农庄。

最初的计划是在文塔河建造一个度假屋，但是在设计的最后阶段，这里变成了一个为社交家庭设计的、靠近库尔迪加的综合住宅建筑群。

这里更适合人生活和居住。

“8 座黑色建筑”组群包括：

1. 围绕着庭院的日常生活建筑群，这也是这组建筑群的核心。

2. 稍远的客房及桑拿房。

3. 依地形而建的地窖和附属建筑。

4. 足球场和排球场。

“8 座黑色建筑”将典型的断面作为所有建筑物的共同元素来使用，将库尔迪加传统的综合住宅建筑群与现代方法相结合。这样的结合要求建筑样式具有继承和归属感，同时建筑本身又能为使用者提供生活所需的舒适感。

湍急的河流、起伏的地形和风构成了这里的场地特征，也定义了建筑的极简表现形式。这一项目通过将一种材料同时应用于墙壁和屋顶，实现了建筑外观的统一。建筑外表皮的纹理和细节是由原木铺板实现的。

与主体的 8 座黑色建筑形成鲜明对比的是，内院与白色聚碳酸酯建造的车库相连，在视觉上为内院增亮。

“8 座黑色建筑”是由 8 座黑色油漆木板建筑构成的建筑组群。

建筑的大窗户建立了与周围环境的视觉联系。不同尺寸的阳台也为住户提供了与自然环境接触的机会。室内空间遵循了嵌套原理，用内置家具墙分隔出一个个独立的房间。阁楼也被建造为额外的就寝空间。

以通道和阳台与其他居住建筑分开的两层建筑，一楼是壁炉室，二楼是业主工作室。这座建筑的墙壁构造与其他建筑不同，是在金属框架上安装了玻璃填充物和稀有木质壁板。壁板可以向外打开，以便住户观赏文塔河的景色。

地下室依地形而建。地下室坚实的岩石墙也成为花园围墙的一部分，还能保护花园免受盛行风的影响。地下室远离生活区，但又通过车库入口与生活区建立了视觉联系。建筑组群还有一间门房，并设置有宽阔、舒适的主入口，以及犬舍和技术设备室，在视觉上打造私人庭院的效果。

建筑组群主入口的墙上设置有特色标识，这是将总平面图变形后制作的。未来主浴室的白墙上也会绘制这样一幅经过叠加、组合而构成的绘图。

Opposite, above: Aerial view of the site. Opposite, below: View of the main building and tower with library from the southeast.

对页，上：鸟瞰图；对页，下：自东南方看主楼及带有图书室的塔楼。

NRJA
2 SISTERS
Langstiņi village, Garkalne 2009

NRJA
姊妹屋
加尔卡尔内，郎斯蒂村　2009　02

Credits and Data
Project title: 2 SISTERS
Program: Private house for two families
Location: Langstiņi, District of Riga, Latvia
Project year: 2007
Built: 2009
Architect: NRJA
Project team: Uldis Luksevics, Martins Rusins
Structural engineer: Aigars Fridrihsons
General contractor: Abi Buve
Plot area: 1,530 m²
House area: 335.4 m²

The building consists of two independent residential premises, based on reinforced concrete pilotis, forming covered external spaces.

The first floor is a connecting space for the entrance zone and rooms of common use.

The way these spaces are connected allows the creation of individual living environments and maximum isolation with a view to Langstiņi Lake.

这座建筑由两个独立的住宅楼组成。它以钢筋混凝土桩柱为基础，形成有遮蔽的开放空间。

一楼连接了入口和两栋楼的公共房间。

这样的连接方式能够创造出各自独立的生活环境，同时每栋楼都可以最大限度地欣赏到朗斯蒂湖的美景。

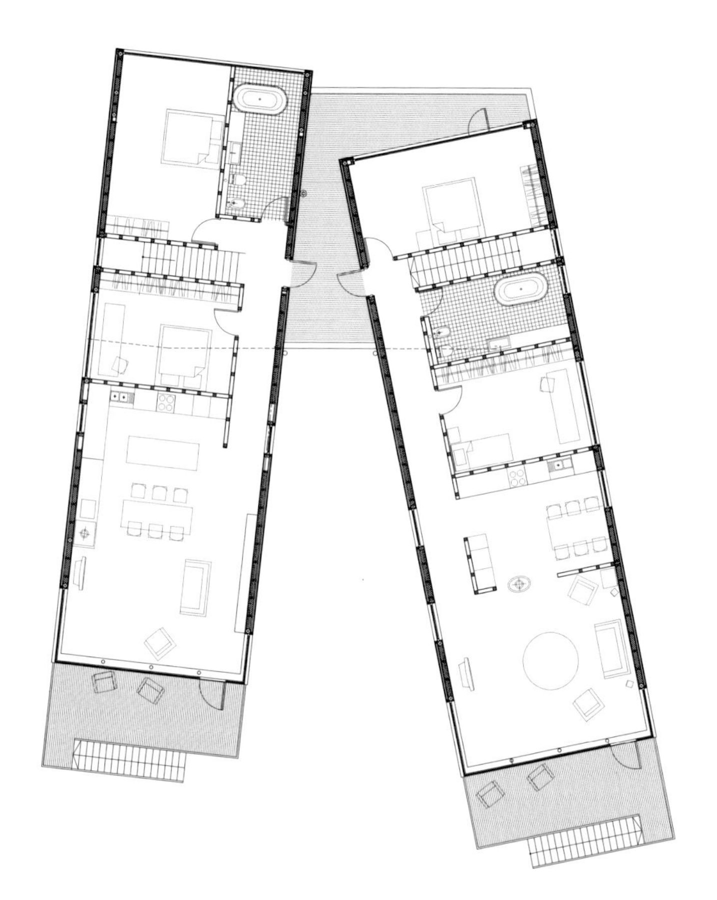

Second floor plan (scale: 1/250)／二层平面图（比例：1/250）

pp. 224-225: General view from the south. Opposite: View of the interval between the two volumes. All photos on pp.224-227 by Gatis Rozenfelds, courtesy of the architect.

第 224-225 页：自南面看到的全景。对页：两栋建筑之间。

NRJA

House of Ruins

Saka parish, Pāvilosta 2006

NRJA

废墟小屋

帕维洛斯塔，萨卡教区　2006

03

The House of Ruins is located in Latvia on the coast of the Baltic Sea. Saka is a rural area. It is the is the least populated district in Latvia, only 2 people per km², because of the large forest reservation (Grinu reservation) and the shore meadows nearby. More and more the area is becoming a vacation / second home area for people from the large cities looking for solitude, privacy and shelter from the big city rush.

It is a new family house built inside the 19th-century ruins of a traditional Latvian barn. The architects here have used the idea of contrast where wind from the sea is opposed to the warmth of the family, and the perfection of glass is set against the rough surface of the old stone. The house provides both the comfort of modern life and the quietness of the nature. Organized in one level, it also contains a small courtyard and a spacious roof terrace for watching the sea and surrounding meadows.

These were not precious ruins. The owner and architects felt that it would be better to preserve the memory of previous time – ruins. The ruins are about 300 m from the sea. They serve as a shelter against offshore wind, but the idea was to create a contrast between:

– the existing weighty walls and the glazed new building,

– the correct forms (concrete walls) of the new building and the existing crumbling wall,

– the rural landscape meadow and the lawn inside the ruins.

Credits and Data

Project title: House of Ruins
Location: Saka parish, Pāvilosta
Project year: 2002
Year completed: 2006
Architect: NRJA
Project team: Uldis Luksevics, Martins Osans
Structural engineer: A222
Contractor: RBS Skals
Area: 200 m²

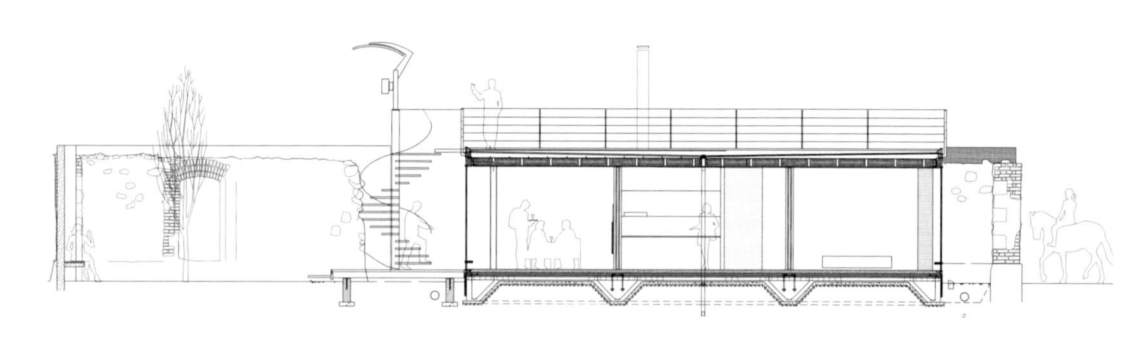

Longitudinal section (scale: 1/200)／纵向剖面图（比例：1/200）

p. 229: View of the interval between the existing wall and new structure. This page: General view from the south side. All photos on pp. 228–231 by courtesy of NRJA.

第 229 页：原有墙壁和新结构之间的空间。本页：南侧全景。

废墟小屋位于拉脱维亚波罗的海沿岸。萨卡地处乡村地带，由于存在大面积的森林保护区（格林努保护区）和海滨牧场，这里也是拉脱维亚人口密度最小的地区，仅有 2 人 / 平方千米。不过近年来，这里逐渐成为都市居民逃离城市喧嚣，寻求独处、隐秘和宁静的度假村或第二居所。

废墟小屋是建造在一座 19 世纪拉脱维亚传统谷仓废墟内的新家庭住宅。在设计上，建筑师采用了对比的理念，将来自海洋的风与住宅的温暖、光滑的玻璃与古老而粗糙的岩石对比。这座住宅既有现代生活的舒适，又有大自然的宁静。此外，这座一层建筑还设有小中庭以及可以眺望海与牧场的宽敞屋顶天台。

这座废墟并没有格外珍贵的价值或魅力。但业主和建筑师都选择保持废墟原貌，留下过去的记忆。这座废墟距离大海 300 米，可以遮蔽离岸风。同时，这座小屋希望在以下几个方面形成对比：

- 现有的厚重墙壁和镶有玻璃的新建筑；
- 新建筑整齐的混凝土墙和原有的破旧墙壁；
- 田园牧场景观和废墟中的草坪。

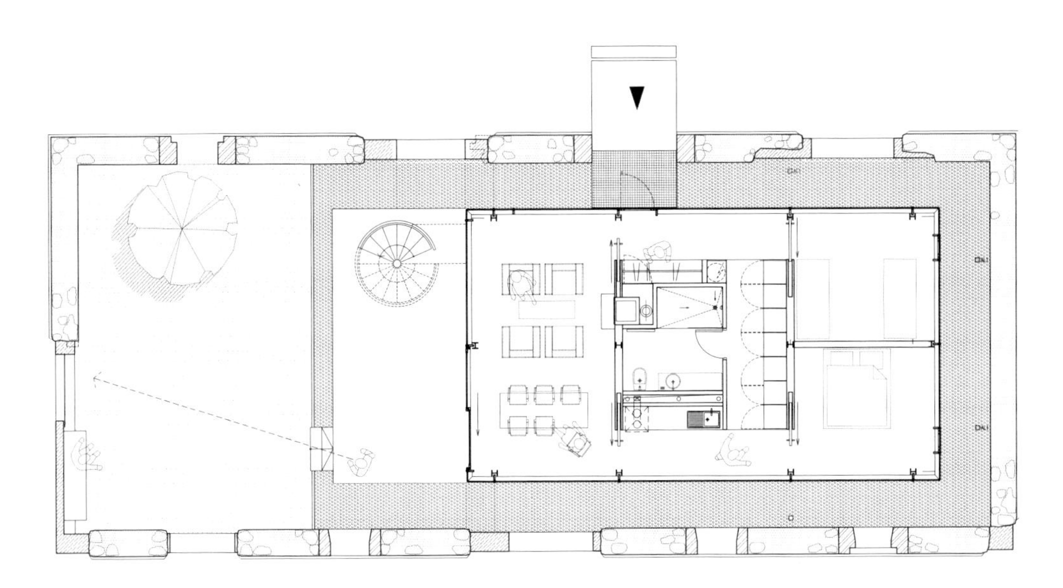

Plan (scale: 1/200)／平面图（比例：1/200）

Zaiga Gaile Office

Vacation Home on Easter Island

Roja parish, Roja 2010

扎格·盖雷建筑师事务所

复活节岛度假屋

罗哈市，罗哈教区　2010　04

On a fine Easter morning in 2005, the architect's family set out on a daytrip along the sea and they found a stone island in the bay of the Baltic Sea with ruins of a building. It turned out to be a former fish farm pumping station, built in the 1980s. Over the 20 years since it was built, it had never been exploited. This monument of Soviet industrial architecture was transformed into a vacation home for the architect's family. The vacation home was registered as Easter Island. The original Easter Island in the Pacific Ocean was also discovered at Easter. To celebrate the legend the images of the stone Moai are perforated in the sliding iron shutters of the house.

The aim of the project was to preserve the island's landscape and the architectural features of the building, the volume and color of which had become a local landmark over the years.

The facades of the building were covered with corten steel plates and shutters to keep the original reddish tone. New large windows were introduced on both sides of the house – facing the sea and the land. The building is no longer a massive block; it has become open and transparent. In the center of the main facades there are glass doors in the same dimensions as the windows. The windows and the

doors are outfitted with sliding shutters. The house can be thus opened or closed to the sun, the wind and the cold. The velvety-brown rust color and the simple shape of the building achieve a delicate harmony with the picturesque coastal landscape.

The rectangle plan of the building is divided into three parts – in the center there is a huge hall with lounge and kitchen area with ceiling as high as 7m. A huge fireplace is installed in the center of the lounge area. One end of the building houses a two-level master apartment, the other end four two-level apartments for children or guests, each with a separate bathroom. The layout of the house

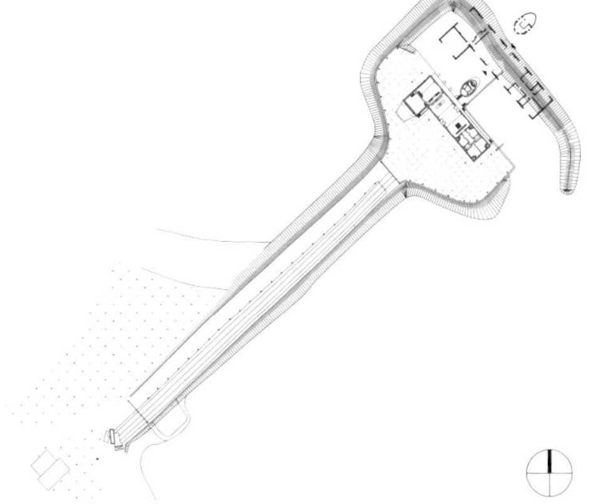

Site plan (scale: 1/4,000)／总平面图（比例：1/4,000）

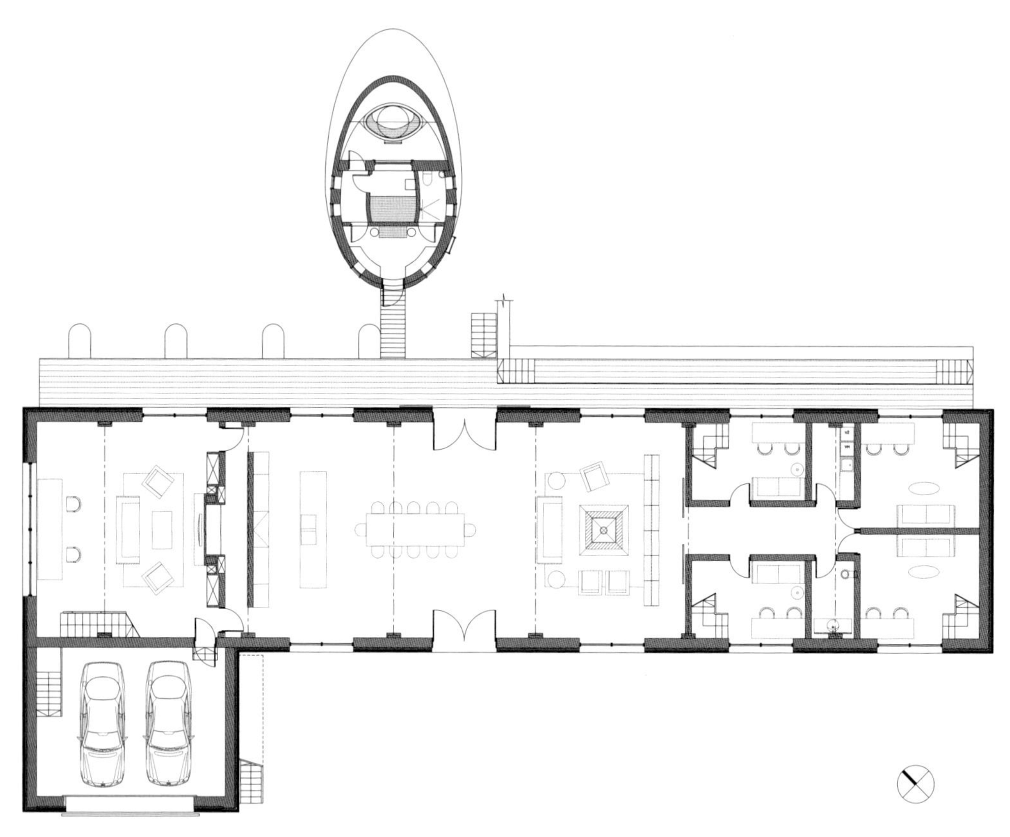

Ground floor plan (scale: 1/300)／一层平面图（比例：1/300）

is simple and symmetrical, as dictated by the clear architecture and the simplicity of the horizontal lines of the landscape – just the sea, the sky and the shore.

The interior design concept was determined by the industrial origin of the building. It is an exercise in contrasts: rough, cool industrial surfaces and perfectly finished, warm wood. The stark element is represented by the crudely finished brick walls, the concrete panels, the metal beams, platforms, stairs and railings, the aluminium windows. The contrast is achieved with consistent and abundant use of wood – a style borrowed from the architecture of yachts and ships. Inside the house, all wooden details are made of exotic Brazilian jatoba cherry wood. The few furniture pieces in the house are carefully selected design icons that have stood the test of time.

The exterior architectural details such as the gates and the outside lamps are also made of rusted steel. They look as if they have been there forever, corroded over the years.

Credits and Data

Project title: Vacation Home on Easter Island
Location: Village of Kaltene, Roja municipality, Latvia
Completed: 2010
Architects: Zaiga Gaile, Agnese Sirmā
Contractor: Māris Gailis
Area of the site: 10,067 m²
Developed area: 442 m²
Pumping station area: Living space 400 m², basement 322 m², total floor space 798 m²
Bathhouse floor space: 52 m²

PARDOD

2005 年晴朗的复活节早上，建筑师的家人去波罗的海海边一日游。他们在海湾发现了一座石岛，岛上保留有一座废弃房屋。那是一座建于 20 世纪 80 年代的养鱼场抽水站，但是在过去的 20 多年中它从来没被使用过。随后，这座苏联时期的工业建筑被改造为建筑师一家的度假屋。它被登记为“复活节岛”，因为太平洋上的复活节岛也是在复活节时被发现的。为了纪念这一逸闻，度假屋的铁质滑动百叶窗上还打孔出摩艾石像的样子。

这一改造项目的目的是保护岛上的景观以及建筑物的特色。在长期的发展中，这座房屋的体量和颜色已经成为当地的标志。

建筑外立面覆盖着柯尔顿钢板和百叶窗，并保留着原有的红色色调。建筑师在面向海洋和陆地的两个方向安装了大窗户，使得整个建筑不再坚硬、冰冷，反而具有开放、通透的特征。在主立面的中央有一扇与窗户等大的玻璃门。门窗上都安装有滑动式的百叶窗，可以依据光照、气温及风向情况开合。建筑丝绒般的铁锈色以及简朴的造型与如画的海岸景观相协调。

这座矩形建筑可以分为三部分。中部屋顶高达 7 米，设置有带休息区和厨房的巨大大厅。休息区中央有一个巨大的壁炉。建筑一端是两层高的业主公寓，另一端是四个供儿童或客人使用的两层公寓，每栋公寓都有独立浴室。受建筑风格以及由海、空、海岸构成的简单地平线的影响，整座建筑的布局简洁而对称。

小屋的内部装修与这座建筑的产业背景密切相关，它在粗犷、冰冷的工业外观和柔和的木材之间营造了一种对比。房屋鲜明简约的特征通过粗制的砖墙、混凝土板、金属梁以及平台、楼梯、扶手、铝窗表现出来。与之对比的是屋内大量木材的使用，这是从船舶的构造中得到的启发。屋内所有的木构细节都是用具有异国情调的巴西加托巴樱桃木制成的。为数不多的几件家具也经过了精心挑选，经得起时间的考验。

诸如大门、外部照明等建筑外部细节是用生锈钢材制作的。这让它们看起来像是经过了时间的洗礼而被腐蚀，仿佛一直都矗立在那里。

名为“鹦鹉螺”的全新雕塑式浴室位于这座历史性建筑的后面，能够让人联想到潜水艇的炮塔。其表面被铆接的不锈钢板所覆盖。

这座房屋在供暖、通风、供水和排污多方面都是可持续的。

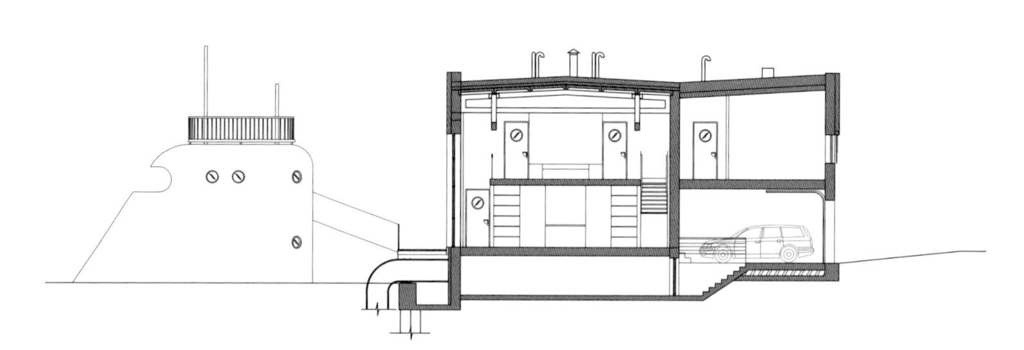

Cross section (scale: 1/300)／横向剖面图（比例：1/300）

pp. 232–233: General view from the southeast. Photo by Ansis Starks, courtesy of the architect. p. 235, above: View from the southeast. Two photos by Sergeis Kondras. p. 235, below: The original building. Photo by Māris Gailis, courtesy of the architect. Opposite: Interior view of the dining room with kitchen.

第 232-233 页：自东南方向看到的全景。第 235 页，上：自东南方向看；第 235 页，下：改造前的建筑。对页：餐厅及厨房内景。

MADE arhitekti

Saldus Music and Art School

Saldus town, Saldus 2013

MADE建筑师事务所

萨尔杜斯音乐与艺术学院

萨尔杜斯市，萨尔杜斯镇　2013

The building of the Music and Art School comprises two schools working separately until now. The classrooms are placed on the perimeter, with practicing halls and libraries in the middle of the building. Light courtyards are the result of the compact plan, providing a lot of daylight and reflected light in the middle of the school, and at the same time being a space for both schools to interact. The color green in the interior marks the Music school, while blue is for the Art school.

The large thermal inertia of the building and integrated floor heating deliver an even temperature regime.

The facade, consisting of massive timber panels covered with profile glass, is a part of the energy efficient natural ventilation system, preheating inlet air during winter. Massive wood walls with lime plaster accumulate humidity, providing a good climate for people as well as for musical instruments inside the classrooms. The building structure and materials work as passive environmental control at the same time as exhibiting its functionality. Inner concrete walls and the massive outside wood wall visible through the glass exhibit their natural origin, which we believe is an important issue especially at institutions of education. There is not a single painted surface on the facade of the school building – every material shares its natural color and texture.

Credits and Data

Project title: Saldus Music and Art School
Client: Regional Municipality of Saldus
Location: Avotu street 12a, Saldus, Latvia
Project: 2007–2009
Construction: 2011–2013
Architect: MADE arhitekti
Project team: Miķelis Putrāms, Linda Krūmiņa, Evelīna Ozola, Uldis Sedlovs,
Liena Amoliņa
Graphic design: Zigmunds Lapsa
Area: 4,179 m² (including courtyards 339 m²)

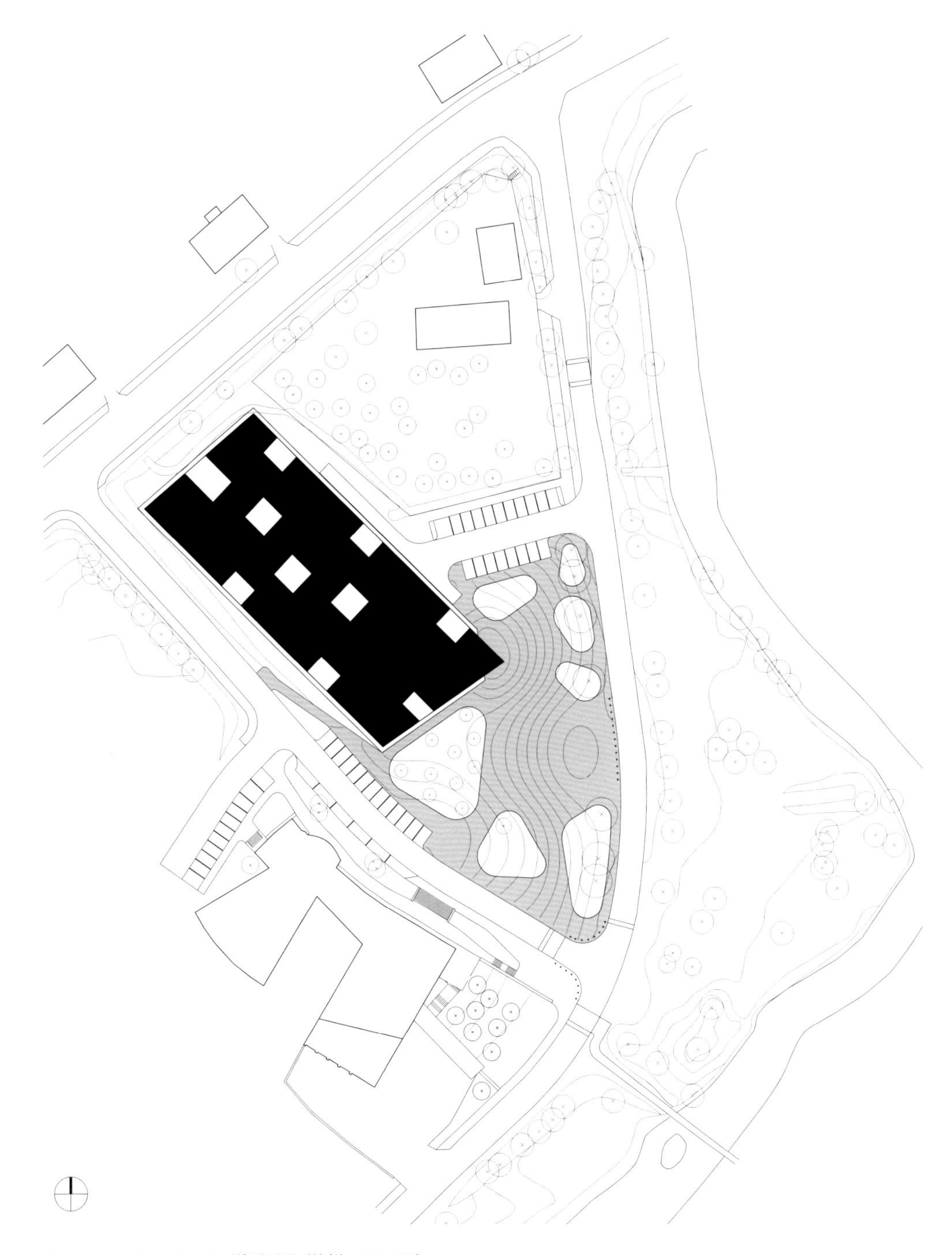

Site plan (scale: 1/1,500)／总平面图（比例：1/1,500）

pp. 238–239: Close up view of the double-skin facad, which consists of massive timber panels, covered with profile glass. Opposite: General view from the southeast. p. 243: View of a corridor. All photos on pp. 238–243 by Ansis Starks, courtesy of the architect.

第 238-239 页：双层表皮立面的特写。其由厚木板构成，覆盖型材玻璃。对页：自东南方看到的全景。第 243 页：走廊。

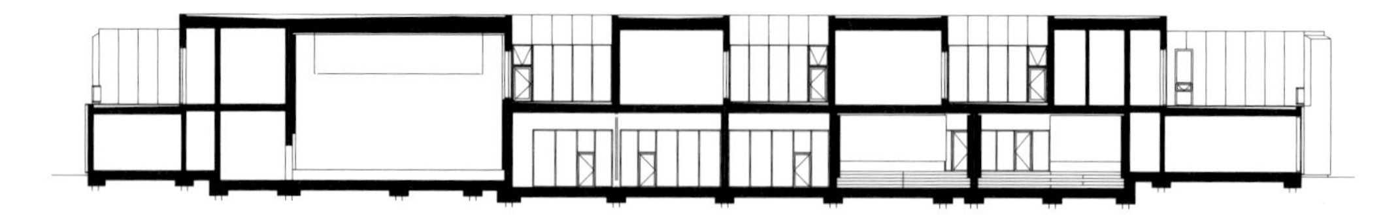

Section (scale: 1/600)／剖面图（比例：1/600）

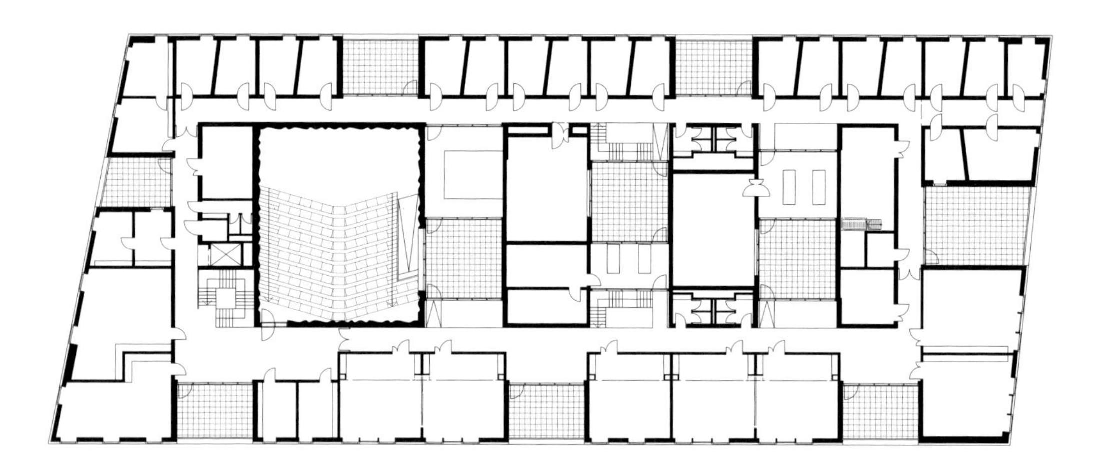

Second floor plan／二层平面图

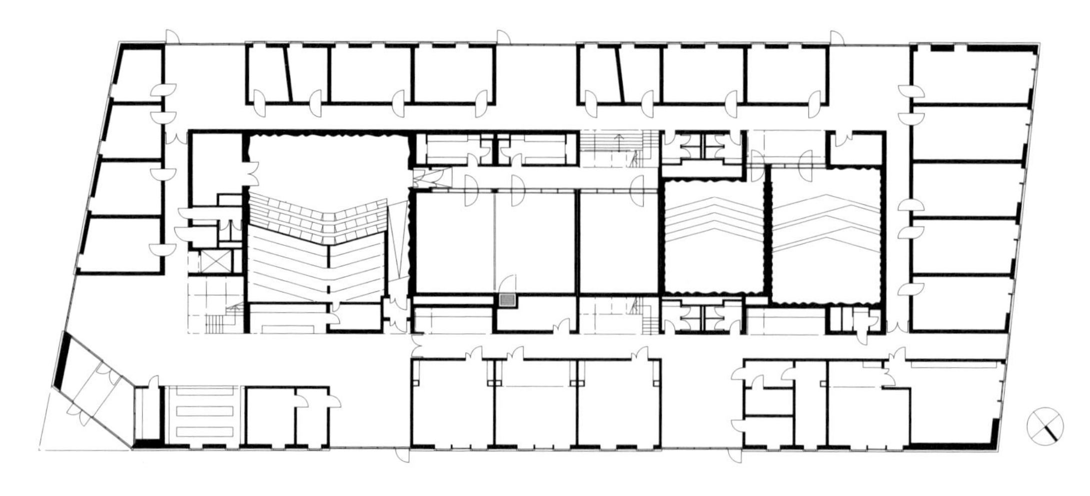

Ground floor plan (scale: 1/600)／一层平面图（比例：1/600）

音乐艺术学院现由两家独立运营的学校组成。教室被设置在建筑物的外围，而练习室和图书馆位于建筑中央。出于紧凑设计的需要，光之庭院能够为建筑内部带来丰富的阳光和反射光，同时也能成为两所学院的互动空间。绿色的室内空间代表音乐学院，蓝色为艺术学院。

建筑内部较大的热惯性和集成式地暖，方便学院进行稳定的温度管理。

建筑外表皮由覆盖有型材玻璃的厚木板构成。这也是节能自然通风系统的一部分，可以在冬季预热进气。用石灰灰泥砌成的厚木墙能够蓄积湿度，为教室内的人和乐器提供良好的环境。建筑结构和材料在辅助被动环境控制的同时，也具有功能性。建筑内部的混凝土墙和透过玻璃可见的厚木板墙是自然材料。这是一个重要因素，尤其是对教育机构来说。此外，建筑外表皮没有经过涂装，每种材料都展示出其自然的颜色和纹理。

Didzis Jaunzems, Jaunromāns un Ābele

View Terrace and Pavilion

Krievkalna island, Koknese parish, Koknese 2013

迪兹·詹泽姆建筑师事务所，尧恩姆斯与伯勒建筑师事务所

观景休憩亭

科克内斯，科克内斯教区，克里夫卡尔纳岛 2013

06

View Terrace and Pavilion are situated in The Garden of Destiny memorial park, in the Consolation area, which is the first zone of Future according to the overall project of the island. The Garden of Destiny is a memorial place for all souls that have been lost to Latvia in the last century and it will be completed as a gift to the country on its 100th birthday in the year 2018. The View Terrace project started as an architectural competition and with the help of donations is now the first permanent building to be realized in the memorial park. The tight bound between the Latvian people and nature has been emphasized in the project. Nature is a source of inner energy, strength, peace and harmony and consolation for Latvians. The project has been designed considering and using particularities of the sights – trees, relief, the most stunning view points. The Viewing Terrace and building have diversified levels of "openness". This creates the opportunity to use the building under all kinds of weather conditions as well as lets visitors choose the level which suits them best. The volume of the pavilion is designed so that it gradually grows from a bench into the building. The building is a platform for harmonious interaction between people and nature.

Credits and Data

Project title: View Terrace and Pavilion
Location: The Garden of Destiny memorial park, Krievkalna Island, Koknese, Latvia
Year: 2013
Architects: Didzis Jaunzems, Laura Laudere in collaboration with architecture office Jaunromāns un Ābele
Awards: Latvian Architecture Award 2013

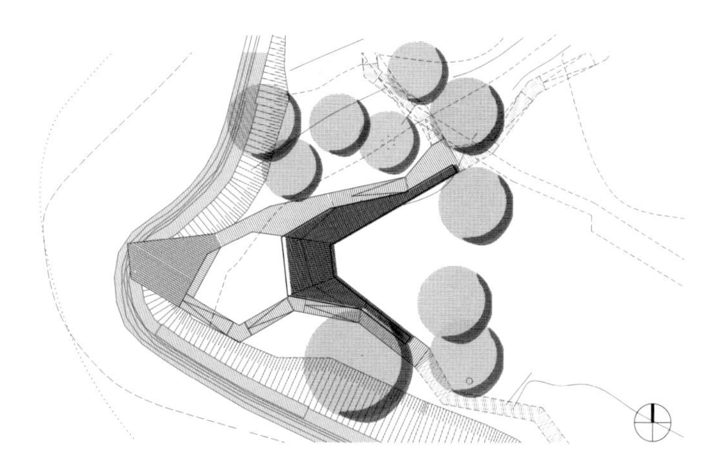

Site plan (scale: 1/600)／总平面图（比例：1/600）

pp. 244–245: General view from the east. All photo on 244–249 by Maris Lapins, courtesy of the architect.

第 244-245 页：自东面看到的全景。

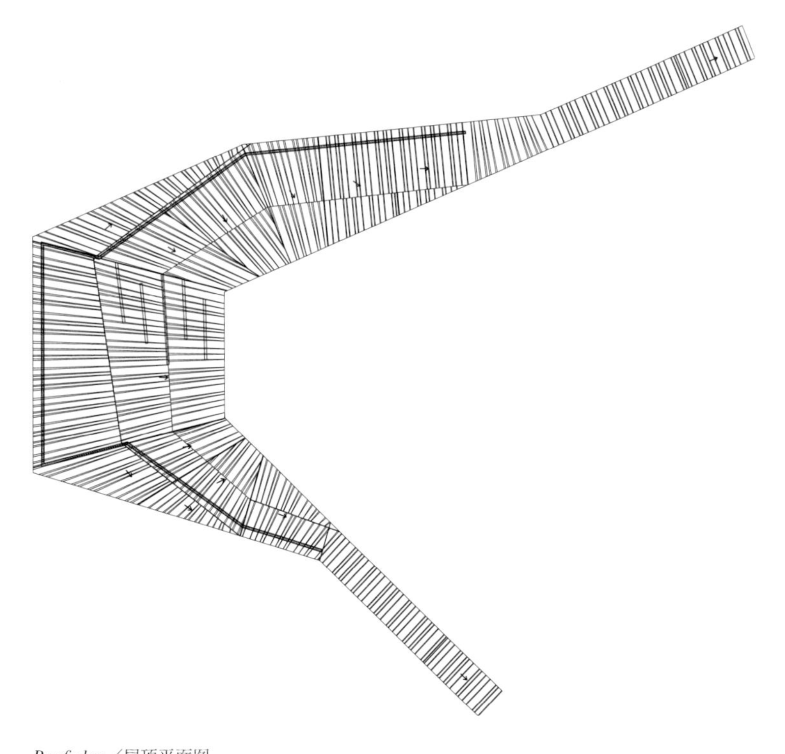

Roof plan／屋顶平面图

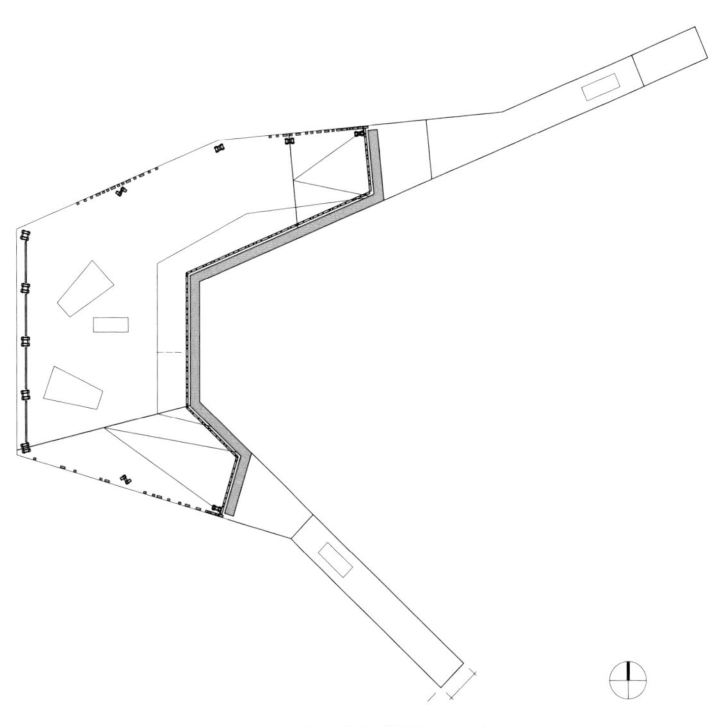

Ground floor plan (scale: 1/150)／一层平面图（比例：1/150）

观景休憩亭位于“命运花园”纪念公园内。依据克里夫卡尔纳岛的整体规划，这是“未来区”的第一站，也是休息区。命运花园是为纪念所有在 20 世纪为拉脱维亚献出生命的人而建造的，于 2018 年拉脱维亚建国 100 周年之际完成。在募捐之下，观景亭项目始于建筑竞赛，现已成为纪念公园内第一座永久性建筑。这一项目突出了拉脱维亚人与自然的紧密联系。大自然是拉脱维亚人内在能量、平和、和谐以及安逸生活的源泉。而这一项目充分考虑了景观的特殊性，将树木、地形等令人印象深刻的景观加以活用。观景休憩亭还具有多层次的“开放性”，能够让游客自主选择观景高度，并适应不同的天气条件。休憩亭的体量由长椅逐渐过渡为建筑。它是人与自然和谐互动的场所。

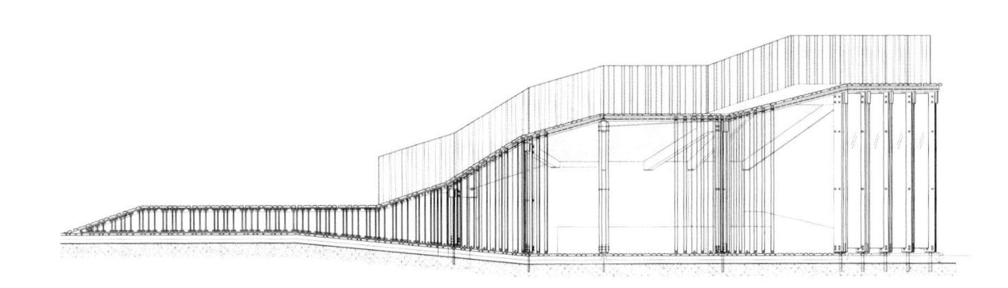

North elevation (scale: 1/150)／北立面图（比例：1/150）

Opposite: View of interior benches. The terrace has two viewing spaces for viewing from different heights.

对页：室内长椅。观景亭具有两个不同高度的眺望空间。

Mailītis A.I.I.M. + Architect J. Poga Office

Latvian National Open Air Stage

Mežaparks neighbourhood, Riga 2007-2022

梅尔蒂斯建筑师事务所+波加建筑师事务所

拉脱维亚国家露天舞台

里加，米亚帕克斯社区 2007-2022

07

The Latvian National Open Air Stage is located in the north part of Riga in Latvia. The stage is integral part of the Song and Dance Festival which is a unique feature of Latvian culture and a part of national identity. The key function of an open air stage is to keep the tradition of the Song Festival, which is included in the UNESCO culture inheritance list. The architects propose to extend the exiting Open Air Stage that was built in 1956 so that more singers and a larger audience can participate in the festival.

The artistic and architectonic concept of the National Open Air Stage is based on the following set of symbols: the Hill, the Tree, the Grove, the Leaves and the Leafage. These symbols comprise the general image and scenic frame of the stage, which is associated with the Latvian landscape, themes from Latvian folk songs and the wisdom of life. The singers' tribune symbolically represents the Hill. The roof construction is based on the idea of a Tree. The interlaced trees make a Grove – the construction of the roof, overlaying an open stage and the singers' tribune. Special manipulated devices are attached to the roof construction – acoustic shields, sound equipment, screens, light decorations, rain protection elements, etc. This changeable part of the stage gives an opportunity to adjust the construction to different arrangements with appropriate acoustics and scenography.

The reconstruction solution basically keeps the existing planning structure of the block of buildings intact. The stage itself gets a new geometry, its capacity increases up to 11,200 places for singers, 23,900 seats and 57,400 standing places for the audience. A two-level open zone for shops and restaurants will be built under the audience area.

The aim of the reconstruction is to turn the stage into multifunctional complex that can be used for different public performances all year round.

Credits and Data

Project title: Latvian National Open Air Stage
Client : Property Department of the Riga City Council
Location: Mežaparks, Riga, Latvia
Year: 2007–2022
Architects: Mailītis A.I.I.M. + Architect J. Poga Office
Project team: Austris Mailītis, Ivars Mailītis, Matiss Mailītis
Area: 10,800 m²

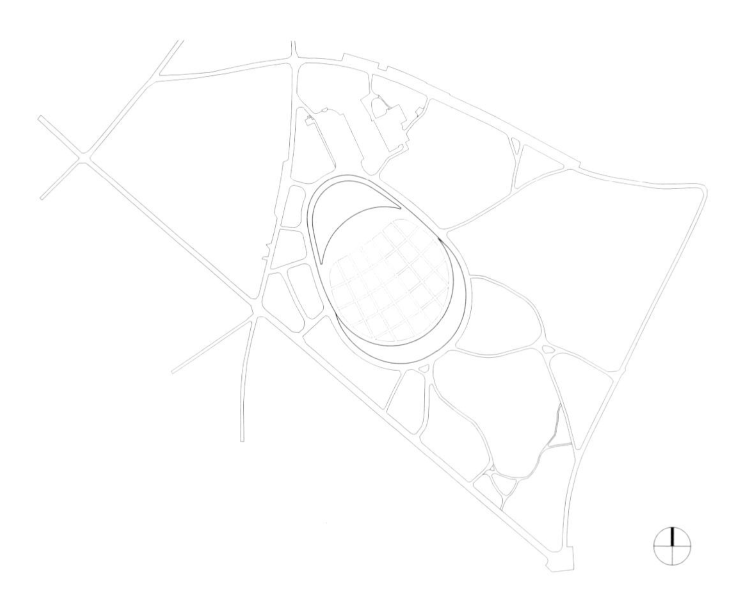

Site plan (scale: 1/12,000)／总平面图（比例：1/12,000）

pp. 250–251: Image of the stage at the Song Festival. p. 254: Image of general bird's-eye view.

第 250-251 页：歌谣节的舞台效果。第 254 页：全景鸟瞰效果。

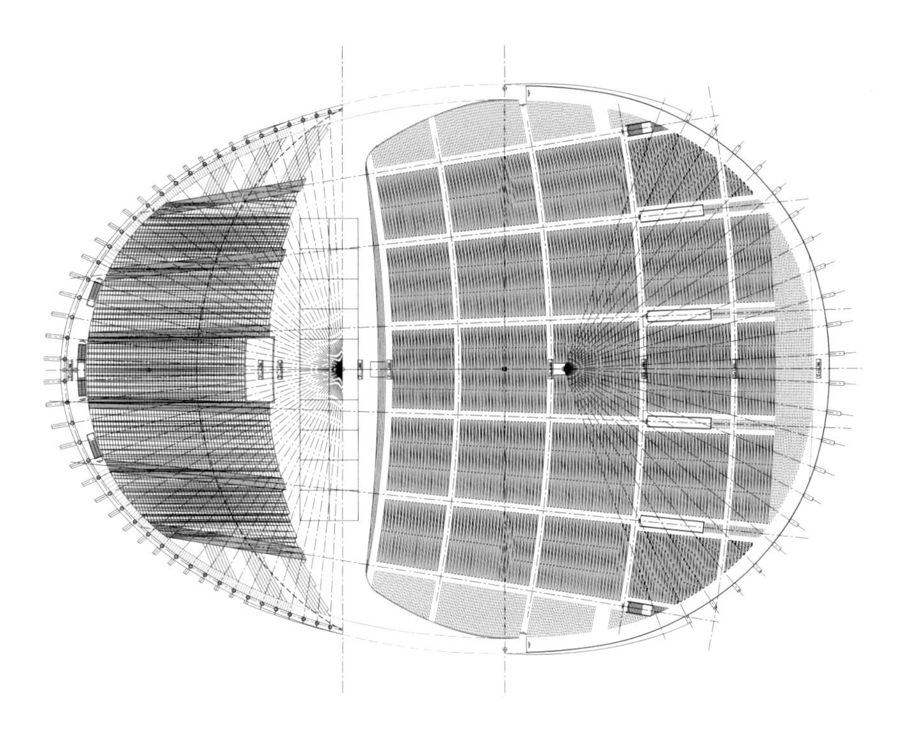

Second floor plan／二层平面图

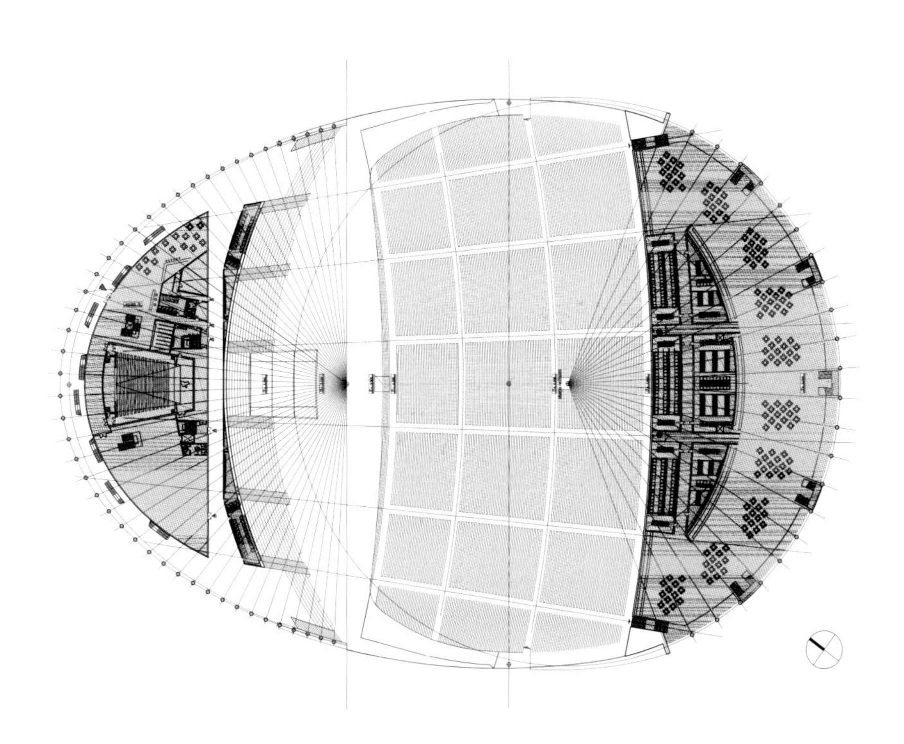

Ground floor plan (scale: 1/2,000)／一层平面图（比例：1/2,000）

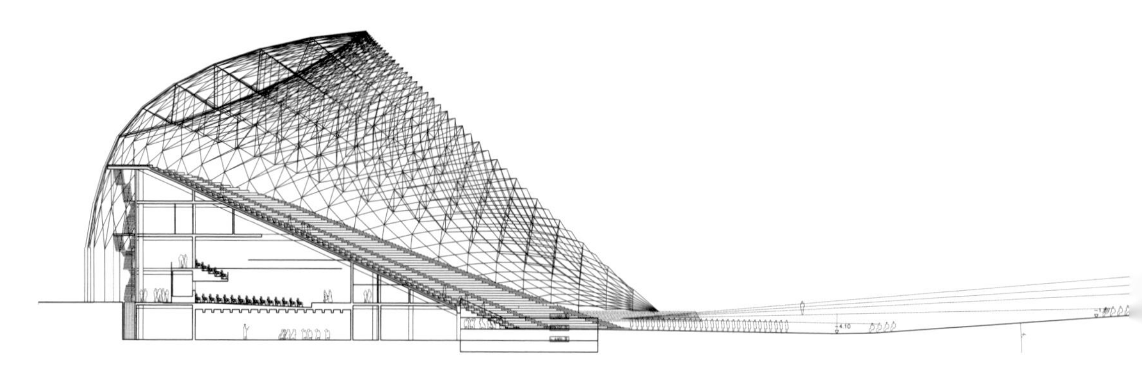

Section (scale: 1/1,000)／剖面图（比例：1/1,000）

拉脱维亚国家露天舞台位于里加北部，是为歌谣节而建造的。歌谣节是拉脱维亚文化中一个独特的要素，也是拉脱维亚民族认同的重要组成，已被列入联合国教科文组织文化遗产名录。而露天舞台的主要功能就是举办歌谣节。因此，建筑师建议扩建 1956 年建造的室外舞台，从而能够让更多歌手和观众参加这一音乐节。

拉脱维亚国家露天舞台的艺术和建筑概念基于山丘、树木、树林、树叶和叶丛等要素。它们构成了舞台的总体形象及视觉框架，既与拉脱维亚景观相和谐，也与这里的民谣主题和生活智慧相关。歌手的唱台象征着山丘。屋顶构造以树木为灵感，向舞台及唱台延伸。而交错的树木就构成了小树林。屋顶上还安装了隔音板、音响设备、屏幕、灯光装饰、防雨装置等特制的控制装置。这样的设计在保证音响和视觉效果的同时，也方便人们依据不同的安排调整机构。

改造方案基本上保留了该地块现有规划的原貌。不过，舞台本身获得了新的几何结构，其容量也增加为歌手席 11,200 个，观众座席 23,900 个以及观众站席 57,400 个。观众区下方将建成两层高的开放区域，用来布局商铺和餐厅。

改造的目的是将这一舞台变为一个多功能综合设施，以适应全年不同类型公共演出的需要。

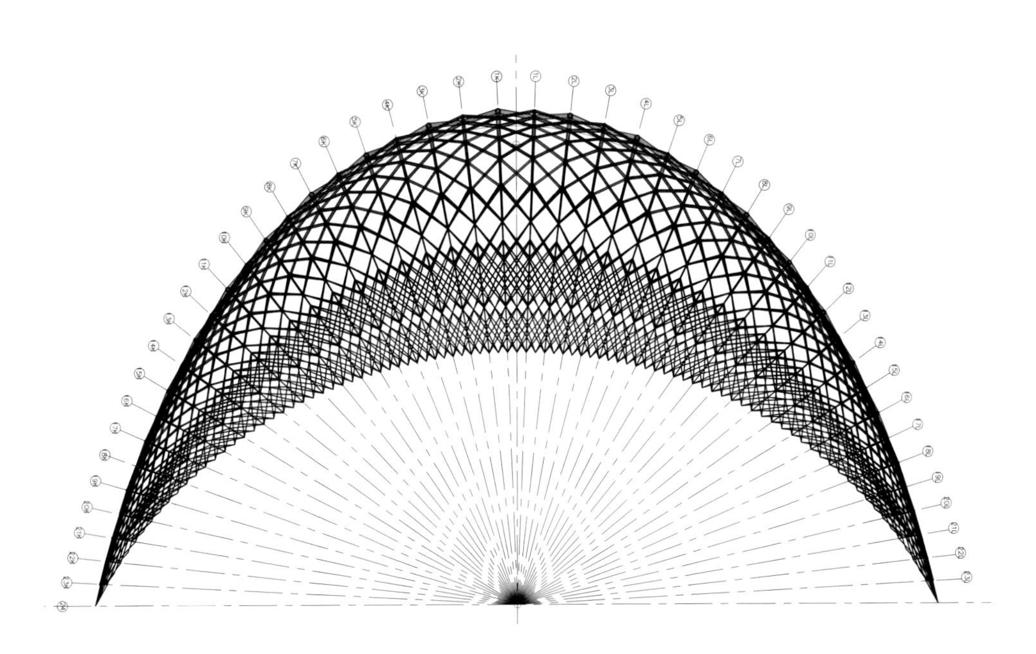

Drawing of roof geometry (scale: 1/1,200)／屋面轴线图（比例：1/1,200）

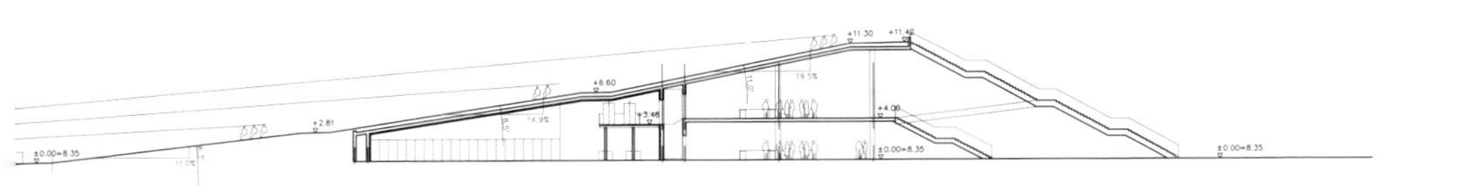

Interview 2:

Coexistence of old and new values in architecture

Interviewer: Ilze Paklone

访谈 2

建筑新旧价值的共存

采访者：伊尔泽·帕克罗内

How would you characterize your own practice? What is important in your work and what factors affect it the most?

I have always liked reconstruction. I remember that during my studies I got acquainted with the issue of Architectural Record addressing the theme of reconstruction and was fascinated by it. Already my diploma project was related to urban intervention – reconstruction of the urban block at the very heart of the historical center of Riga. Unfolding both our tangible and intangible cultural legacies underlay all my work. A careful attitude towards historical substance and its changes, finding balance between heritage and contemporary interventions are very important for me. My work always starts with meticulous studies of history of the place, understanding authenticity and structure of historical layers.

Zaiga Gaile, Zaiga Gaile Office

I can say that Latvia is a very pure, unspoilt and remarkably balanced place. The activities of architects can be very dangerous as we can transform the environment swiftly and radically. Therefore I feel that an underlying sense of wanting to understand the context, structural organization, spatial relationships and scaling of elements related to it and eventually rationality have been important aspects of our practice. Context also implies that the design is in perfect alliance with the programs that it serves. Rationality towards designing has always been something distinct to Latvians. There is a lot to learn from this pragmatic rationality.

Someone once said that we are in some way contextualists. It seems to me that contextualism gives the opportunity for our designs not to be simply marvelous objects but rather to gain a lot of their value from the existent physical context and through new interpretation or innovation in it. One of the valuable aspects of contextual design is that it is not in extreme interference in the existent reality. It rather goes beyond it and thus has the potential to become timeless value. Such design is not selfish in the sense that it does not demand attention by any cost. It is vitally important that the design almost seamlessly blends into the environment as the self-evident part of it and most importantly creates the space around it with a particular value. The external form is relevant, but we rather create poetic imagery from the interconnections between context, program, usability and values.

Andris Kronbergs and Raimonds Saulītis, Arhis

It feels that this is the right moment to take care of the old buildings. "To take care of" is the best expression to describe my work of reconstruction, renovation and restoration. My creativity manifests itself in establishing connections between the old and the new. The imaginative moment can be found even in locating the technical innovations in the old buildings. That is the pleasure to feel that something old performs on a new level yet

您如何总结自己的实践特色？工作中您最看重什么，最大的影响因素是什么？

我一直很喜欢重建。我记得我在上学时曾接触到一期以重建为主题的杂志《建筑实录》，并为之着迷。我的毕业作品也与城市介入有关，即在里加老城中心重建城市街区。而彻底了解本地的物质和非物质文化遗产，是所有工作的基础。对我来说，谨慎对待历史遗存及其变化，找到遗产和当代介入之间的平衡，都相当重要。在开始工作之前，我总会对当地历史进行细致考察，理解历史的原真和构成。

——扎格·盖雷，扎格·盖雷建筑师事务所

我可以说，拉脱维亚是个十分纯净、淳朴、表现均衡的国度。建筑师的活动在某种情况下显得极其危险，因为它能够快速、彻底地改变环境。我认为我们的实践中很重要的一点，是在潜意识里愿意了解建筑相关的文脉、结构形态、空间关系、要素尺度，并在此基础上保持理性。对脉络的关注也意味着设计能够完美契合项目的需求。拉脱维亚人一直清楚，应当理性对待设计。而这种务实的理性也有许多借鉴意义。

有人曾说，我们在某种意义上是文脉主义者。在我看来，脉络主义为我们提供了机会，让我们的设计不仅仅是让人惊异的物体，而是能够通过其他阐释或融合创新，从现有的物理背景中汲取价值。基于文脉来设计的一个可贵之处便是，不会过度介入现状。它更多是超越现实，从而有可能产生永恒的价值。这样的设计决不能以自我为中心，乃至不计代价博得关注，它至关重要的是与环境融为一体；而比这更重要的，是为周围空间创造出特定的价值。外在形式固然有意义，但我们更倾向联结脉络、项目、实用性和价值，并由此创造美好的图景。

——安德里斯·克朗博格和雷蒙兹·萨乌提斯，
阿尔希斯建筑师事务所

我感觉现在正是呵护老建筑的最佳时机。“呵护”最适合用来形容我这些重建、改造、修复的工作。我可以通过为新旧事物建立关联，证明自己的创新力；即使只是对老建筑进行一些技术变革，也会有灵光乍现的瞬间。我很乐于让旧的事物焕然一新，同时尽量隐藏技术或建筑上的介入。对于西欧而言，呵护老建筑是难得的介入物质遗产的机会。拉脱维亚同样如此。

——雷尼斯·列平修，苏德拉巴建筑师事务所

关于拉脱维亚建筑界和我们自己的实践变化，如果结合建筑师的职业来谈，可以总结出三个方面。首先，在新的建筑实践中（我们也是践行者），不再单独给某位建筑师贴上“大师”标签，团队里每个专业人士都是独立且平等的，层次关系更加扁平化。这还体现在工作流程和最终结果上。就连事务所的名字，也少有以某个人名命名的做法了。其次，年轻一代的建筑师对于周围环境有着强烈的责任感和敏锐的意识，他们明白任何行动或项目都会以一定形式影响更大的环境，能够认识到自己的影响力和关联性。第三，沟通模式已经改变。我们一直在亲身实践中

the technological and architectural interventions are almost invisible. It is very special opportunity of dealing with physical heritage for the Western Europe. That includes Latvia.

Reinis Liepiņš, Sudraba Arhitektūra

If we talk in relation to the profession of architects, we could summarize the changes of architectural discipline in Latvia and also our own practice in three main directions. Firstly, new architectural practices, including our own, are collectives or unions of independent and equal professionals without one architect that could be labelled as the master. Relationships are more horizontal. Also the process of organizing the work and the end product. It manifests itself even in the names of the offices. There are relatively few that are named after one person. Secondly, the young generation has a strong sense of responsibility and awareness of surrounding processes. It is evidenced in the understanding of how any action or the components of the project will affect the broader context. It is comprehension of one's own influence and connections. Thirdly, the mode of communication has changed. The aspect of communicating important social and cultural manifestations through architectural means is something we have been studying in our practice. It is the matter of unfolding architecture as a democratic process where many participants are involved rather than as an incomprehensible discipline exclusive only to few. The profession has extended its boundaries to understanding that architecture is not only about physical building, but also about theoretical, social or any other modes of communication.

Evelīna Ozola and Toms Kokins, Fine Young Urbanists

At this moment we actually are two offices. Didzis Jaunzems Architecture deals with theoretical researches and interests, small-scale objects, scenographies, installations and environmental objects. The other, MAAS, manages relatively larger scale architectural projects and takes care of participating in competitions. My work experience at OMA and professional studies impressed on me the need to think about architecture as research, vision and narrative. A particular subject that I have been interested in lies in exploring scenography as the potential to design unique spatial situations, both short term and long term, that affect viewer's perception. Tearing down boundaries of understanding and disciplines is a very special feeling. For instance, we are used to permanency in the urban environment – solid buildings, fixed streets. Scenographic elements have the potential to make us seeing the mundane city in a surprising manner.

Didzis Jaunzems, Didzis Jaunzems Architecture

研究，通过建筑手段反映重要的社会和文化表现。建筑应当是大众可以参与的民主过程，而非只有少数人能理解的学科。建筑师的职业边界已经进一步外扩，使人意识到建筑不仅是物理上的房屋，还与理论、社交以及各种类型的沟通手段有关。

——艾薇莉娜·奥佐拉和汤姆斯·科金斯，
都市青年建筑事务所

我们目前实际上是两家事务所。迪兹·詹泽姆建筑师事务所进行理论研究，关注小型对象、场景布置、装置、环境对象；MAAS事务所负责相对大规模的建筑项目，并参与设计竞赛。在OMA的工作经历和专业学习让我意识到，有必要把建筑当作研究领域、愿景或叙事。在了解场景布置时，我对设计独特的空间情境尤其感兴趣，它可以在长时间或短时间内影响观众的感知。突破认知和学科之间的边界，总会带来一种十分特殊的感受。举个例子，我们习惯了城市环境的恒定，例如坚固的建筑和固定的街道，但通过场景布置，日常生活的城市有可能以一种出人意料的方式展现在我们面前。

——迪兹·詹泽姆，迪兹·詹泽姆建筑师事务所

Could you describe the context of designing just after Latvia regained its independence in 1991 and how the situation has evolved during the last 25 years? What were the priorities then and at the current moment? What do you think architecture in Latvia has accumulated during these 25 years?

Over these 25 years my attitude towards my work and my focus on reconstructions has not changed. Despite the pressure to build faster and cheaper, I consistently do what I consider important – sensible attitude towards the context while adjusting it to contemporaneity. I do not have ambitions to build something that can be evidently noticed. Subtle dialog between historical substance and its changes, how you twist them so that their perform in a new way is more important for me. This stance can be seen in my works in Ķīpsala Wooden Collection (See pp. 184–209), Gypsum Factory (See pp. 172–177), Berga Bazaar (See pp. 64–73) or Rumene Manor.

Rather, the world has changed. I feel that now there is more care towards the heritage. It can also be seen in the couple of last Venice Biennales as an overall tendency – no showing off with expressive forms, no visualizations. Work with the heritage is the given starting terrain for the architects in Latvia.

Zaiga Gaile, Zaiga Gaile Office

If we talk specifically about the evolution of thinking over the last 25 years, it has rather been continuous thinking about the same themes and context over and over again than a salutatory changes of ideas.

The office was founded a little before independence was regained. In principle, initiation and evolution of the office complied with the processes related to cultivating the new situation. In principle, everything has changed since then. Until independence, many of the large-scale important buildings in the city had not been developed. After 25 years we are not isolated from the rest of the world as we were, thus new designs are not radical marvels in the local cultural scene anymore. At this moment integration, collaborations, careful studies of materials, sustainability issues have come to the foreground and are bigger challenges instead of creating new surprises. Twenty five years ago designing was more about getting to explore and experiment with unknown terrains, the things we did not know. That is something that has changed a lot. The attitude towards historical values has changed, allowing more space for coexistence of the old and new values. There is no longer strictly dogmatic approach. There is an opportunity to discuss construction of the values, to estimate them.

All the actualities in cultural processes reflect with certain deviance and delay in architecture as design needs to intended, though over, drawn and implemented. It is rather difficult to grasp in the exact moment when something is happening, but the consequences and interconnections can be comprehended over time. Awareness towards the surroundings has been growing towards the

请您描述一下在1991年拉脱维亚恢复独立之后，设计的总体环境和最近25年的发展状况。当时和现在的重点分别是什么？您认为拉脱维亚的建筑在这25年间获得了哪些积累？

25年来，我对本职工作的态度以及重建的关注从没变过。尽管更快、更廉价的建造要求令人倍感压力，我始终在做自己认为重要的事情：保持对脉络的敏感，同时积极适应时代。我没有那种建造引人注目的东西的野心，对我而言更重要的是历史遗存及其变化之间的均衡对话，以及如何用新的形式展现它们。这从我过往的作品中也能感受到，例如基普萨拉岛上的木建筑群（见184-209页）、石膏工厂（见172-177页）、贝格斯集市（见64-73页）或鲁麦尼庄园。

但世界变化了。我感觉现在遗产得到了更多关注，过去两届的威尼斯双年展也显示出这样的倾向。建筑不再炫耀夸张的外观和视觉效果。对于拉脱维亚的建筑师来说，他们更是一开始就需要与遗产打交道。

——扎格·盖雷，扎格·盖雷建筑师事务所

价值得到了新的对待，新旧价值有了更多共存的空间。循规蹈矩的方法被摒弃，我们有机会讨论如何构建价值、评估价值。

文化进程会投射到建筑上，但往往存在偏差和延时，因为设计需要确定意图，继而思考、绘制、实施。尽管很难抓住事件发生的瞬间，但随着时间的推移，也能理解事件的结果以及相互之间的关联。人们对周围环境的认识，正在逐渐向着理性的一面发展，少了吹嘘夸耀，多了对事物本质的追求。这对建筑师而言至关重要，毕竟我们的主要任务之一就是理解真实的需求，而不是那些不切实际的愿望。

或许，我们已经迎来“小项目、高质量”的时代了。

——安德里斯·克朗博格和雷蒙兹·萨乌提斯，
阿尔希斯建筑师事务所

具体说到过去25年思想的变迁，比起思想上的飞跃，更多是关于同一主题、同一脉络的反复思索。

我们的事务所成立于恢复独立前不久。原则上，事务所的启动和发展都紧随着新的形势，所有事物也从那时起发生着变化。但直到拉脱维亚独立，这座城市许多重要的大型建筑项目都仍未动工。25年后，我们不再与世隔绝，新的设计也不再是本土文化景象中的奇观。整合、协作、材料研究、可持续发展等需求都已走到台前，这些与其说是制造着惊喜，不如说是更大的挑战。25年前，设计更多是对未知领域的探索和实验，但今天的情况大为不同。历史

1990年代充斥着困惑，毕竟我们朝着未知跨进了一大步，试图更快、更专业地应对变化。尽管我们历来是西欧文化圈的一部分，但在恢复独立之后，我们显然更想强调这一点。我们在西欧文化圈中的位置可以说是一个意味深远的历史结果，这也反映在当下重建里加城堡的项目中。10世纪初，东征十字军到达这里，想要压制本土的异教文化，从而开启了我们的西欧文化环境。久而久之，我们总体倾向欧洲文化，但同时也与斯拉夫文化保持着相对亲密的关联。

过去，苏联与西方世界之间存在着一道想象中的界线，即所谓“铁幕”。我们曾经将幕后的事物过度理想化，认

side of reasonableness – less ostentation and more appreciation of the essence of things. This aspect is vital importance to architects as is one of our tasks to focus attention to understanding real needs instead of unreasonable wishes.

Perhaps, now is the time of smaller projects with higher architectural qualities.

Andris Kronbergs and Raimonds Saulītis, Arhis

A lot of confusion was present in 1990's. It was a huge leap into the unknown and an attempt to react swiftly and professionally to changes. Although historically we were already part of the Western European cultural space, after regaining independence there was a clear desire and vector to emphasize it. Our place in Western European culture is an interesting historical consequence, also in the context of our current project – reconstruction of the Riga Castle castellum. Initiation to the Western European cultural space began with the crusaders arriving here at the early 10th century and their attempts to suppress the local pagan cultures. Over the time our inclination towards the European culture materialized although we have been relatively close also to the Slavic cultures.

We idealized what we thought was behind the so called Iron Curtain, the imaginative partition between the Soviet Union and Western world. We tended to think about the West as something unattainably beautiful. Over these 25 years we have stabilized and have been able to generate our own understanding and beauty.

After regaining independence it made a difference to travel and see architecture that had been only studied from the literature, to see positive as well as negative aspects of it. It took some time to assimilate it into our own thinking. The young generation, in turn, they do not have those boundaries, they are already there by default.

Reinis Liepiņš, Sudraba Arhitektūra

为西方象征着某种不可企及的美好。经过25年的发展，我们终于安定下来，已然能够形成自己的理解和独特的美。

恢复独立之后，通过实地探访那些原本只能通过文献进行研究的建筑、亲身体会建筑的优缺点，可以给我们带来不同的感受。将这些感受融入自己的思考，着实费了些时间。不过年轻一代已经察觉不到这样的边界了，在他们看来那些都是理所当然的。

——雷尼斯·列平修，苏德拉巴建筑师事务所

Architects Profile

建筑师简介

Processoffice is architecture and urbanism practice based in Vilnius, Lithuania. It was founded in 2007 as a continuation of six-year-long collaboration by principal architects Vytautas Biekša (second from left), Rokas Kilčiauskas (left) and Marius Kanevičius (right). Processoffice is aimed at providing specific system for exchange where the network of specialists from various fields is engaged in theoretical and practical issues of architecture and urbanism.

Portrait courtesy of the architect.

Process建筑师事务所位于立陶宛维尔纽斯，专注于建筑及城市化实践。事务所创立于2007年，是主持建筑师维托塔斯·比克萨（左二）、罗卡斯·基利亚乌斯卡斯（左）和马里乌斯·卡内维乌斯（右）长达6年合作的延续。事务所旨在为各领域专家讨论建筑及城市化的理论和实践提供专门平台。

SZK and Partners was established in 1995. Over the past 20 years, the office has implemented more than 50 projects covering a wide range of professional activities from urban planning solutions and hundreds of thousands of square meters of living space, to the design of furniture and light fixture for public and private interiors. The notable projects include the Riga Concert Hall, the extension of the Art Academy of Latvia, Mercedes Benz centre in Riga, several apartment buildings and private houses in Jurmala and Riga.

Portrait courtesy of the architect.

SZK合伙人建筑师事务所创立于1995年。在过去的20年中，事务所已经实施了超过50个项目，其范围涵盖了从城市规划方案及数千平方米的生活空间设计，到公共及私人空间内的家具和灯具设计。其著名项目包括里加音乐厅、拉脱维亚艺术学院扩建、里加的梅赛德斯–奔驰中心、尤尔马拉和里加的几栋公寓楼和私人住宅。

Fine Young Urbanists started in 2011 as an informal collaboration between Evelīna Ozola (right) and Toms Kokins, and was officially registered as a design studio in 2013. Works include renovation of Alekša Square in Riga (2014), ephemeral installation Mierīgi! in Riga (2014), and remodelling of Līvu Square in Cēsis (2015). In their projects, Fine Young Urbanists employ proactive methods of communication to the public through architecture and involve local communities in the design process.

Portrait by Kaspars Kursišs, courtesy of the architect.

都市青年建筑事务所始于2011年，最初源自伊芙娜·奥佐拉（右）和汤姆斯·科金斯之间的非正式合作，随后在2013年正式注册为设计事务所。他们做过的项目包括里加亚历山大广场改造（2014年）、里加米埃里基！临时装置（2014年）以及瑟斯鲁广场改造（2015年）。在项目中，他们善于利用建筑与公众展开积极对话，将当地居民纳入到设计的过程当中。

Zaiga Gaile (born in 1951) is a Latvian architect, an expert in the renovation and conversion of historic architecture, the author of numerous books and articles, and the owner of her own architectural firm **Zaiga Gaile Office** since 1992. She has been awarded numerous prizes including the Grand Prix of the Latvian Architecture Award 2014 for the reconstruction of Riga School of Design and Art (2013), and nomination for Mies van der Rohe Award for Wooden Building Reconstruction of the Riga School of Design and Art and Žanis Lipke Memorial (2012).

Portrait courtesy of the architect.

扎格·盖雷（1951-）是拉脱维亚建筑师、历史建筑更新与改造专家，同时她还著有大量书籍和文章。1992年，她创立了自己的建筑事务所**扎格·盖雷建筑师事务所**。她获得了众多奖项，例如里加设计与艺术学院重建项目（2013年）获得了2014年拉脱维亚建筑大奖，里加设计与艺术学院木构建筑改造项目及扎尼斯·利普克纪念馆改造项目（2012年）被提名密斯·凡·德·罗奖。

Architect Reinis Liepiņš is a founder of **Sudraba Arhitektūra** and one of those rare experts who are able to combine a fine approach to architectural heritage with the needs of his contemporary clients. Liepiņš was an architect for the Latvian exposition at the Venice Architecture Biennale 2008. He received B.Arch from Faculty of Architecture and Urban Planning, Riga Technical University. Currently the office is working on Renovation Project of the Castellum of Riga Castle (National History Museum of Latvia).

Portrait courtesy of the architect.

建筑师雷尼斯·列平修是**苏德拉巴建筑师事务所**的创始人之一，是少数能将建筑遗产保护与当代客户需求完美结合的专家之一。列平修也是2008年威尼斯建筑双年展拉脱维亚博览会的建筑师之一。他获得了里加科技大学建筑与城市规划系的学士学位。目前，事务所正在筹备里加宫（拉脱维亚国家历史博物馆）的改造项目。

Arhis was founded in December 7, 1988 but started working in February 3, 1989. The extent of its work includes architecture, interior design and engineering projects. Arhis has received many recognitions and prizes such as four times the Latvian Architecture Award of the Year for the Building Reconstruction and Addition in Palasta Street 7 (2000), The Riga International Airport reconstruction 2nd phase (2001), a private residence in Gulbju Street 6, Sigulda (2004) and the Printing Office "Britania" in Daugmale (2010).

Portrait courtesy of the architect.

阿尔希斯建筑师事务所创立于1988年12月7日，正式运营开始于1989年2月3日。工作范围涵盖建筑设计、室内设计和工程项目。事务所也获得了众多荣誉和奖项，如曾四次荣获拉脱维亚年度建筑改造奖，获奖项目分别是帕拉斯塔大街7号改造（2000年）、里加国际机场二期重建（2001年）、古尔布朱街6号私人住宅改造（2004年）以及道格马尔印刷厂“布里塔尼亚”（2010年）改造。

Jaunromāns un Ābele was established by Māra Ābele (left) and Mārtiņš Jaunromāns in 2005. Ābele studied architecture at Riga Technical University, Polytechnic University of Milan and University of Paris 8 in Saint-Denis. Since 2008, she has been a lecturer at Faculty of Architecture and Urban Planning, Riga Technical University. She is also a member of Ethics panel of Latvian Association of Architects and Council of creative industries at Latvian Ministry of Culture. Jaunromāns studied architecture at Riga Technical University and Polytechnic University of Milan.

Portrait courtesy of the architect.

尧恩姆斯与伯勒建筑师事务所由玛雅·伯勒（左）和马蒂斯·尧恩姆斯于2005年创立。伯勒曾在里加科技大学、米兰理工大学和圣丹尼斯的巴黎第八大学学习建筑。自2008年起，她一直担任里加科技大学建筑与城市规划学院的讲师。她还是拉脱维亚建筑师协会伦理小组成员和拉脱维亚文化部创意产业委员会委员。尧恩姆斯曾在里加技术大学和米兰理工大学学习建筑学。

MARK arhitekti was founded in 2006. MARK has a wide range of professional specialization, which includes development of architectural, interior and engineering design, as well as project management. While realizing a project, a special attention is drawn to sustainable construction and BREEAM methodology solutions are offered. Within more than ten years of professional existence, more than 85 designs of different scales – such as multi-storey public buildings, private houses and interiors – have been developed.

Portrait courtesy of the architect.

MARK建筑师事务所创立于2006年。他们的业务范围较广，包含有建筑、室内设计、工程设计以及项目管理。他们在项目中格外看重可持续性建设以及BREEAM方法的使用。在十多年的专业生涯中，他们已完成了85个以上不同规模的设计项目，包括多层公共建筑、私人住宅及室内装饰等。

NRJA is a young Riga-based architectural practice established in 2005 by Uldis Lukševics (born in 1969). Most of its work is obtained through invited competitions where the office aspires to propose more than what is allowed or required. That is what NRJA stands for – No Rules Just Architecture. NRJA is currently involved in proposals such as a high-rise Z Towers in Riga and several large scale multifunctional housing and mixed-use projects. House of Ruins has received several awards including Latvian Architecture Award, Grand Prix (2006) and was nominated for Mies Van der Rohe award.

Portrait courtesy of the architect.

NRJA是一家年轻的里加建筑事务所，由乌迪斯·卢克舍维奇（1969-）于2005年创立。事务所的项目大多来源于邀请赛，而他们在竞赛中往往表现出超前的思考。这也是NRJA的宗旨——建筑没有规则。目前，事务所正在为里加的一座高层Z字形塔，以及几个大型多功能住宅和综合性建筑做提案。而他们的废墟小屋项目已获得了多个奖项，包括拉脱维亚建筑大奖（2006年），同时也被密斯·凡·德·罗奖提名。

MADE arhitekti was founded by Miķelis Putrāms (1980, right) and Linda Krūmiņa (1979) in 2007 in Riga. They work with inventive and rational concepts that come from project's context and particular issues, engaging passionately in projects with complex tasks such as educational buildings, culture public institutions, housing and urban developments. Putrāms and Krūmiņa studied at Faculty of Architecture and Urban Planning, Riga Technical University. Krūmiņa also studied at School of Architecture, The Royal Danish Academy of Fine Arts.

Portrait courtesy of the architect.

MADE建筑师事务所由米利斯·普特尔斯（1980-，右）和琳达·科米亚（1979-）于2007年在里加创立。他们能够从项目背景和特定的问题出发，运用兼具创造性与逻辑性的理念，完成复杂的项目。他们的项目既有教育建筑、文化公共设施及住房，也包括城市规划。普特尔斯和科米亚曾在里加科技大学建筑与城市规划学院学习，科米亚还曾求学于丹麦皇家美术学院建筑学院。

Didzis Jaunzems Architecture was founded by Latvian architect **Didzis Jaunzems** in 2012. Its work is focused on architecture, urban planning and design project development. Jaunzems has studied and gained practical experience of architecture and urban planning in Latvia, Norway and Italy. He has worked at OMA. In 2013 and 2015, Jaunzems received Latvian Architecture Award for View Terrace and Pavilion in the Garden of Destiny and Nature Concert Hall.

Portrait courtesy of the architect.

迪兹·詹泽姆建筑师事务所是拉脱维亚建筑师**迪兹·詹泽姆**于2012年创立的。事务所专注于建筑、城市规划和设计项目开发。詹泽姆曾在拉脱维亚、挪威和意大利学习建筑和城市规划，并积累了一定的实践经验，也曾供职于OMA大都会建筑事务所。他已先后两次获得拉脱维亚建筑奖，分别是命运花园观景台及展馆（2013年）和自然音乐厅（2015年）。

Mailītis A.I.I.M. is an architectural studio based in Riga, established by architect Austris Mailītis (photo) and artist Ivars Mailītis in 2007. Studio is working in fields of architecture, art, scenography and design. Studio has been authors for large scale projects including Latvian Pavilion in Expo 2010 in Shanghai, Shaolin Flying Monk Temple in China and Latvian National Open Air Stage. Mailītis, born in 1984, graduate from Riga Technical University in 2012.

Portrait courtesy of the architect.

梅尔蒂斯建筑师事务所是一家里加本地的建筑事务所，由建筑师奥斯崔斯·梅尔蒂斯和艺术家伊瓦尔斯·梅尔蒂斯于2007年创立。事务所主要从事建筑、艺术、布景等相关设计，已完成众多大型项目，例如2010年上海世博会拉脱维亚馆、河南嵩山少林和尚飞行剧场和拉脱维亚国家露天舞台。梅尔蒂斯出生于1984年，2012年毕业于里加科技大学。

Gunārs Birkerts was born in 1925 in Riga. In 1949, he graduated from the Technical University Stuttgart with the degree of Dipl.-Ing in architecture. Between 1959–1990, he served as Professor of Architecture at University of Michigan. Birkerts worked for Perkins and Will, Eero Saarinen, and Minoru Yamasaki before establishing his own office in Detroit. He is a fellow of renowned institutions including AIA and Latvian Association of Architects. He has received numerous awards such as the Great Medal, Latvian Academy of Sciences (2000) and the Order of the Three Stars, Republic of Latvia (1995).

Portrait courtesy of the architect.

冈斯·伯克茨1925年出生于里加，1949年毕业于斯图加特科技大学，获得建筑学学士学位。1959年至1990年间，他在密歇根大学担任建筑学教授。伯克茨在底特律创立自己的事务所之前，曾为珀金斯和威尔建筑事务所、埃罗·沙里宁以及山崎实（也音译作：雅马萨奇）工作过。他是美国建筑师协会和拉脱维亚建筑师协会等著名机构的成员。他曾获得众多奖项，包括拉脱维亚科学院大奖（2000年）和拉脱维亚共和国三星勋章（1995年）。

Volker Giencke was born in Wolfsberg, Carinthia, Austria, in 1947. He studied architecture and philosophy in Graz and Vienna. He opened Giencke & Company architectural office in Graz in 1981, as well as offices in Klagenfurt / St. Veit, Austria, and Seville in 1990 and Riga in 2004. He is Professor for Architectural Design at the Institute of Design, University of Innsbruck, since 1992, where he founded ./studio3 – the Institute for Experimental Architecture in 2000. His recent works include: Cultural Forum Berlin, Peach Park Shanghai, and 360° Jakomini, Residential – Roof Extension Graz.

Portrait courtesy of the architect.

沃尔克·吉恩克1947年出生于奥地利卡林西亚的沃尔夫斯堡，曾在格拉茨和维也纳学习建筑及哲学。1981年，他在格拉茨创立了吉恩克建筑师事务所，并于1990年在奥地利圣维特克拉根福和西班牙塞维利亚开设事务所，随后又于2004年在里加开设事务所。自1992年起，他在因斯布鲁克大学设计学院担任建筑设计教授，并于2000年在校内创立了实验建筑研究所。他最近的作品包括柏林文化论坛、上海桃园以及格拉茨360°雅克米尼住宅的屋面扩建。

SAALS is a Latvian architect office founded in 2007 by Rasa Kalnina (left). She has received B.Arch in 2002 from Faculty of Architecture and Urban Planning, Riga Technical University. She has worked at several offices in Riga before establishing SAALS. Maris Krumins has worked at SAALS as an architect since 2007. SAALS's completed works include: BĀKA Lighthouse (2011), Rezekne Creative Service Centre – Zeimuls (2014), both in Latvia, and Multifunctional Hall in Malta (2015).

Portrait courtesy of the architect.

SAALS是一家拉脱维亚建筑师事务所，由拉萨·卡尔尼娜（左）于2007年创立。她于2002年获得了里加科技大学建筑与城市规划系的学士学位。在建立SAALS之前，她曾在里加的多家事务所工作过。马里斯·克鲁米斯自2007年起就在SAALS担任建筑师。SAALS已完成的项目包括拉托维亚BKA灯塔（2011年）、拉脱维亚泽尔斯雷泽克创意服务中心（2014年），以及马耳他多功能厅（2015年）。

INDIA is an abbreviation of Intelligent Design Interior and Architecture – i.e. all the main fields of work of our architecture office. It was established in May 2004. INDIA started as a team of three young architects – Pēteris Bajārs (top left), Rūdolfs Jansons (top right), and Inga Stakione – and soon became one of the most promising architecture offices in Latvia. The office has won many awards including the annual Latvian Architecture Award for its Parventa Library in Ventspils.

Portrait courtesy of the architect.

INDIA是Intelligent Design Interior and Architecture的缩写，也隐含了**INDIA**事务所的主要业务领域。INDIA由三位年轻的建筑师佩蒂斯·巴杰尔斯（左上）、鲁道夫·詹森（右上）和英加·斯塔基翁于2004年5月共同创立。**INDIA**事务所很快就成为了拉脱维亚最有潜力的建筑事务所之一，并获得包括拉脱维亚建筑奖（文茨皮尔斯的帕尔文塔图书馆项目）在内的诸多奖项。

Arnis Dimiņš is architect and founding partner of architectural company Substance, based in Riga. Dimiņš received his degree in architecture at Riga Technical University. Before establishing Substance, he worked for the Latvian architectural company Arhis. Dimiņš has received awards for the best design in several Latvian Annual Architecture and Design Prize competitions. His works were shortlisted for the Mies van der Rohe award and the Russian Architecture award ARHIP. His projects have been widely published in international architectural press.

Portrait courtesy of the architect.

阿尼斯·迪米什是里加Substance建筑师事务所的主创建筑师兼创始人。他在里加科技大学获得建筑学学位。在创立Substance建筑师事务所之前，他曾在拉脱维亚的阿尔希斯建筑师事务所工作。迪米什曾多次在拉脱维亚年度建筑和设计竞赛中获得最佳设计奖，作品也曾入围密斯·凡·德·罗奖和俄罗斯ARHIP建筑奖。此外，他的作品也被很多国际建筑出版社发表。

Brigita Bula Architect was established in 2013 and known for architecture and interior design as well as exhibition design and corporate branding. Completed works include: Family House in Cēsis (2014), Boat Centre in Pāvilosta (2012), Taste Latvia fashion stores in four locations in Riga (2008–2013), and Theatre Bar in Riga (2012). Prior to founding her own studio **Brigita Bula** was a project architect at Arhis. In 2012 Bula received the annual Architecture Award from the Latvian Association of Architects for the Holiday House in Kaltene (2012).

Portrait courtesy of the architect.

布吉丽塔·布拉建筑师事务所创立于2013年，专注于建筑和室内设计、展览和品牌设计。事务所已完成的作品包括瑟斯的家族住宅（2014年）、皮罗斯塔的船舶中心（2012年）、里加的四处拉脱维亚品味时装店（2008-2013年）以及里加剧院酒吧（2012年）。在创立自己的事务所之前，布吉丽塔·布拉曾是阿尔希斯建筑师事务所的项目建筑师。2012年，布拉的卡尔特纳度假屋项目获得了拉脱维亚建筑师协会颁发的年度建筑奖（2012年）。

安藤忠雄全集

TADAO ANDO COMPLETE WORK

致敬

中日邦交正常化50周年纪念项目 日本国际交流基金会赞助项目

世纪传奇建筑家 全解建筑世界里的光影挑战

品 5+家具艺术品 1500+摄影作品 500+手绘作品 100+ 论文及安藤故事 10+ 展览及建筑小品

Spotlight:

Ganso Dream World

Tadao Ando Architect & Associates

特别收录：

元祖梦世界

安藤忠雄建筑研究所

元祖梦世界
立即来体验

Located in Qingpu District, Shanghai,Ganso Dream World is an experiential commercial complex for children and youth. The owner GANSO is a confectionery manufacturer from Taiwan, China, who commits to aiding children's education by organizing activities such as factory tours and baking classes. Ganso envisions a place for children to learn with fun hands-on activities. Hence, in response to the client's needs, we designed the building in the eyes of children

Firstly, to improve the massive scale and spatial visibility, simple geometry such as circles, squares, and equilateral triangles, which are straightforward to children, are introduced as the subject of the composition, where a variety of places and spatial imagery is conceived from mutual combinations and gradually connected multicenters.

The main building is divided into three functional blocks: the children-themed *Dream Sky*, the landscape monolithic *Dream Hill*, and the children's accommodation facilities in the northwest.

Dream Sky features four island areas wrapped in soft, curved facade. Each island contains a respective center and areas, where the open space is connected by a three-dimensional void and a curved circular path. The streamline indoor/oudoor transitional space, guides pedastrain movement and stimulates children activities.

We plan to create *Dream Hill* as a living gallery integrated with a semi-outdoor space with skylights in contrast to *Dream Sky* which flows with significant curvature. The open space, formed by the geometric interplay with three square planes axis, creates the courtyard, plaza and terrace which unfold in three dimensions. The accommodation facility designed for children has an L-shaped volume enclosing the atrium of *Dream Hill*, contains guestroom plans for different users, an architectural façade in triangular shape and distinctive opening pattern, making *Dream Hill* a landmark that echoes with *Dream Sky*.

The landscape masterplanning follows the geometric elements of the building with its circle, square and triangle. Integrating the building, blue and green design is developed along with the architectural "terrain design", the children facility shaped to be *a breath of learning experience for children in nature*. We hope the place will become a strong help for the growth of children who symbolize the future of China and the world.

Text by Tadao Ando Architect & Associates

Credits and Data

Project title: Ganso Dream World
Location: Shanghai, China
Design: 2013.6-2015.3
Completion: 2015.3-2019.8
Structure: Reinforced Concrete
Fuction: Commerce, Hotel
Site area: 43,650.00 m^2
Building area: 21,827 m^2
Total floor area: 148,452 m^2

世界
体验

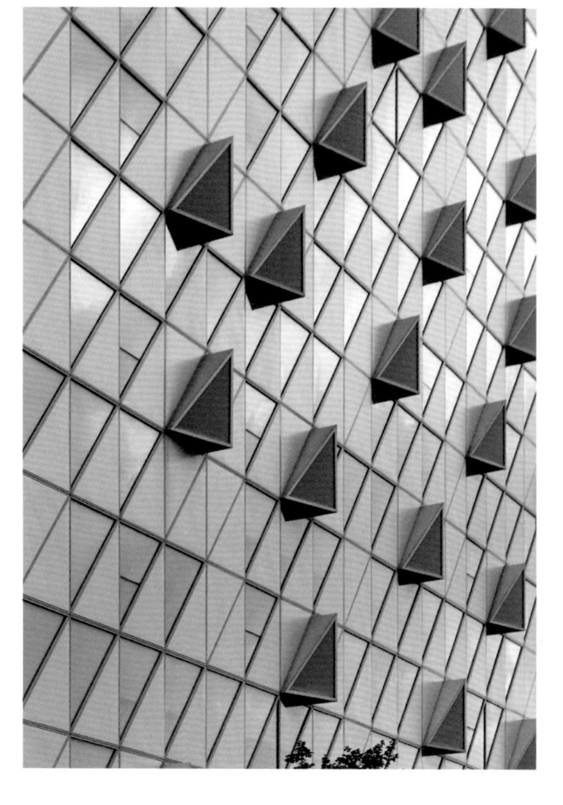

基地位于上海市青浦区，是一座面向少年儿童的体验型综合商业设施。业主是中国台湾的点心制造企业“元祖食品”，他们致力于通过组织孩童参观工厂、体验蛋糕制作等教育活动来助力儿童教育。业主希望能创造出一个让孩子们身体力行的学习体验场所。为了回应这样的需求，我们以孩子的视野为主题进行了设计。

首先，为了解决巨大的规模与空间可视性的问题，我们将圆形、正方形、等边三角形这类孩子容易理解的纯粹几何学作为构图的主题，基于相互组合，创造出各种各样的场所，并构想了多个中心缓缓相连的空间意象。

建筑整体分为三大功能各异的区块：儿童主题的“梦幻天空”街区、与景观一体化的“梦幻山丘”，以及基地西北部的儿童住宿设施。

“梦幻天空”以舒缓的曲面表皮包裹起四个岛状区域。每个岛拥有不同的中心和分区，其中留白区由立体挑空与形成曲线的环状道路相连。内外互动的流线型空间，在引导人们移动的同时，也诱发了儿童的活动。

我们计划将“梦幻山丘”打造为一个与带有天窗的半室外空间合为一体的生活画廊。与描画曲线的“梦幻天空”形成对照。基于几何学的交错所产生的留白，以三个正方形平面为轴线，构成了立体展开的庭院、广场和露台。为儿童而设计的住宿设施，拥有仿佛要围合起“梦幻山丘”中庭的L形体量，内含对应不同使用者的多种客室平面，且建筑外立面以三角形为主题、开口处图案别具特色，使之成为了一座呼应“梦幻天空”的建筑地标。

基地整体的景观规划，也与建筑一样沿袭了圆形、正方形、正三角形这三种几何学主题元素，基于与建筑融为一体从而展开水与绿的设计、及建筑化的“地形”设计，我们希望这座儿童设施整体可成为“在大自然中的自由呼吸的儿童学习体验场”。希望这个场所，能够有力帮助到肩负中国及全世界未来的孩子们的成长。

安藤忠雄建筑研究所/文

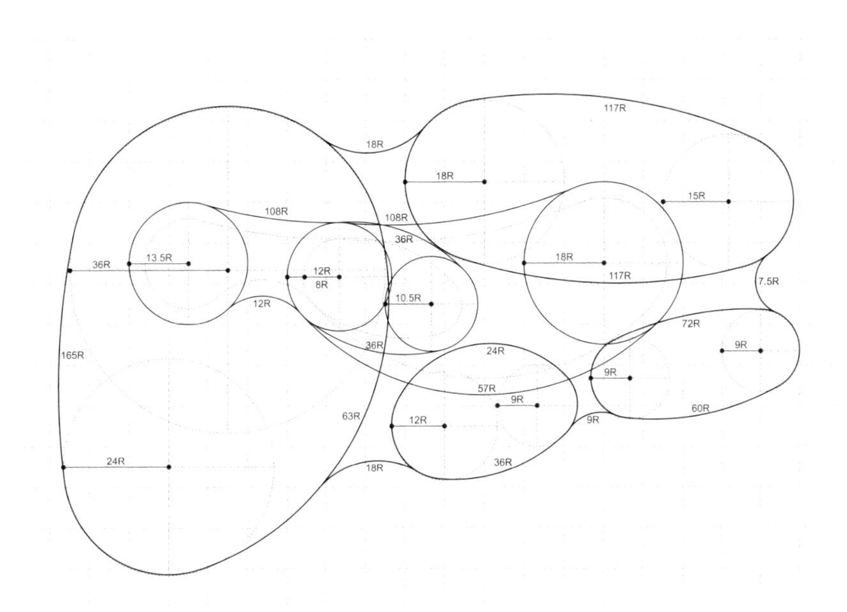

pp. 272-273: The appearance on the road side. p274: The undulating facade of the building. p275: Sketch by Tadao Ando. Opposite, above: The glass corridor connects two different "paradise". Opposite, bottom left: The appearance of aluminum material like ruffled Silk on children's skirt. Opposite, bottom right: Facade of the "Dream House". All images on pp. 272–279 by Luz Images.

第272-273页：临道路侧的建筑外观。第274页：波浪般起伏的建筑外立面。第275页：安藤忠雄手绘稿。对页，上：玻璃连廊连接起了两座不同的“乐园”；对页，左下：像儿童裙花边褶皱一般的铝材质的建筑外表皮；对页，右下：亲子酒店的外立面局部。

Opposite: Concrete columns run the length of the internal streamlined space. This page, above: Soft and dynamic interior space of the building. This page, bottom: Sketch by Tadao Ando.

对页：混凝土立柱贯穿整个内部流线形空间。本页，上：柔和而富有动感的建筑内部空间；本页，下：安藤忠雄手绘稿。